SpringerBriefs in Environmental Science

More information about this series at http://www.springer.com/series/8868

Jyotish Prakash Basu

Climate Change Adaptation and Forest Dependent Communities

An Analytical Perspective of Different
Agro-Climatic Regions of West Bengal, India

 Springer

Jyotish Prakash Basu
Professor of Economics, Former Head, Department of Economics
Former Vice Chancellor
West Bengal State University, Barasat, North 24 Parganas
Kolkata, West Bengal, India

ISSN 2191-5547   ISSN 2191-5555   (electronic)
SpringerBriefs in Environmental Science
ISBN 978-3-319-52324-8   ISBN 978-3-319-52325-5   (eBook)
DOI 10.1007/978-3-319-52325-5

Library of Congress Control Number: 2017936198

Printed on acid-free paper

This Springer imprint is published by Springer Nature
The registered company is Springer International Publishing AG
The registered company address is: Gewerbestrasse 11, 6330 Cham, Switzerland

Contents

Introduction

Climate change is one of the major threats to sustainable development because of its effects on health, infrastructure, agriculture, food security, and forest ecosystems (IPCC 2007a). In an era of global climate change, forests need to be managed in such a way that they play multifunctional roles like contributing to the mitigation of climate change through carbon storage and improving the livelihoods of forest-dependent people (Adams et al. 2004; Benayas et al. 2009; Canadell and Raupach 2008; Chazdon 2008; Lamb et al. 2005; Chhatre and Agrawal 2009; Pandey 2002). Forest management interventions that result in carbon emission reductions or increased carbon sequestration could potentially be rewarded by REDD+ (FAO 2010). Forest is the national wealth of India. Forest cover in India (2013 assessment) is 69.79 million ha which is 21.23% of the geographical area of the country. The tree cover of the country is estimated to be 9.13 million ha which is 2.78% of the geographical area. The total forest and tree cover of the country is 78.92 million ha which is 24.01% of the geographical area of the country.

In India 700 million rural populations directly depend on climate-sensitive sectors like agriculture, forest, and fisheries. Forest ecosystems provide a wide range of economic and social benefits, such as employment, forest products, and protection of cultural values (FAO 2006). Forest-dependent people comprise a significant proportion of the communities most vulnerable to the impacts of climate change on forests. Forest-dependent people are considered society's poorest on the basis of social indicators such as life expectancy, maternal mortality, education, nutrition, and health (ADB 2009).

In India there are around 1.73 lakh villages located in and around forests (MoEF 2006). The forest-dependent population in the country varies from 275 million (World Bank 2006) to 350–400 million (MoEF 2009), most of which are underprivileged populations living in forested regions (Saha and Guru 2003). Forest is considered the most vulnerable sector and constitutes an integral part of the social lives of disadvantaged communities living in and around forest areas, and contributes substantially to the food supply and livelihood security of tribal populations in India.

Thus about 300 million rural poor are dependent on Indian forest and the availability of non-timber forest products like food, fuelwood, medicine, sal leaves, kendu leaves, and mushrooms. The forest dwellers and farmers identified by Byron and Arnold (1999) are particularly at risk due to climate change. The impact of climate change on forest-dependent communities has been documented in a range of countries such as Bhutan (Tshering 2003), Vietnam (Trieu 2003), India (Sharma 2003), China (Shougong et al. 2003), Malawi (Fisher 2004), Mozambique (Lynam et al. 2004), and Ethiopia (Mamo et al. 2007). The vulnerability of many communities in developing countries is immense and their capacity to adapt to the impact of future climate change is often very low (Huq et al. 2004; Mertz et al. 2009a). The presence of adaptive capacity is a necessary condition for the design and implementation of effective adaptation strategies so as to reduce the livelihood and the magnitude of harmful outcomes resulting from climate change (Brooks and Adger 2003). The Intergovernmental Panel on Climate Change (IPCC)'s Third Assessment Report (AR3) supposes that the main features of a community's adaptive capacity are economic resources, infrastructure technology, infrastructure, information and skills, institutions, and equity (IPCC 2001). Studies carried out after AR3 led the Fourth Assessment Report (AR4) to acknowledge the influence of social factors such as human capital and governance structures (IPCC 2007b). The fundamental goal of adaptation strategies is the reduction of the vulnerabilities to climate-induced change.

Annual mean surface air temperature is projected to increase between 3.5 and 4.3 °C by 2100 (MoEF 2012). The climate-induced impacts, along with changes in the frequency and intensity of extreme events such as floods and droughts, are likely to impact human health, agriculture, water resources, natural ecosystems, and biodiversity (MoEF 2012; Srinivasan 2012).

According to the IPCC the vulnerability of a region to climate change depends on its wealth, so poverty limits adaptive capabilities (IPCC 2001). Furthermore, socioeconomic systems are more vulnerable in developing countries where economic and institutional circumstances are unfavorable. Microfinance can reduce the vulnerability of the poor and the possibility of linking this tool to climate change adaptation is of considerable importance (Hammill et al. 2008). The Self Employed Women's Association (SEWA) in India offers housing loans to repair or replace roofs, reinforce walls, or rebuild houses to reduce vulnerability to extreme events such as floods, droughts, and storms (Pantoja 2002). Migration by the poor as a response to natural calamities and other shocks has been documented and is called distress migration (Mukherjee 2001).

A number of indices related to vulnerability, sustainability, and quality of life have gained prominence in literature. Among them are the Environmental Vulnerability Index (Kaly et al. 1999; SOPAC 2005), Environmental Sustainability Index (Esty et al. 2005), Human Development Index (UNDP 1990, 2005), Human Well-being Index (Prescott-Allen 2001), and Prevalent Vulnerability Index (Cardona 2005).

Given the backdrop, the objective of the study sets the following:

1. To identify the main climate-induced vulnerabilities that affect livelihoods of the forest communities, evaluate the expected impacts of climate change at different levels under different climate change scenarios, and assess and measure the implied major social, environmental, and economic impacts of climate change.
2. To assess potential adaptation strategies in the household economy considering economic, social, and environmental perspectives.
3. To examine the factors influencing the adaptation to climate change.
4. To examine the role of governmental policy (including national forest policy of 1988) in reducing climate-related vulnerability.

Hypotheses

1. There is a positive relationship between the education level of the household head and adaptation to climate change (Maddison 2006).
2. Access to information on climate increases chance of adapting to climate change (Yirga 2007; Maddison 2006; and Nhemachena and Hassan 2007).
3. The use of non-timber forest products is an important adaptation strategy.
4. Migration emerged as another important adaptation strategy.
5. Households in the mountain region and coastal region are less likely to take climate change adaptation measures than are households in the drought-prone region in West Bengal.
6. Livestock rearing is also an adaptation strategy.

Chapter 1
Review of Literature

The purpose of this chapter is to make a review of literature on adaptation to climate change and forest dependent communities. This may be analyzed with reference to the following aspects, namely livelihood, vulnerability, adaptation and adaptive capacity.

1.1 Livelihood

A better understanding of the complex nature of livelihoods has derived largely from work conducted in research into poverty (Sen 1981; Narayan et al. 2000). Income measures and ownership of assets like land which failed to capture many key issues of poverty like marginalization and vulnerability. IFAD (2002) considers coastal areas in Asia are prone to poverty and coastal fishing households are regarded as being amongst the poorest of the poor, largely on the basis of their dependence on an open-access resource where competition is high and increasing. The ability of the rural poor to sustain their livelihoods is believed to be constrained due to adverse environmental conditions—high ecological vulnerability and low resource productivity—and limited access to land and other natural resources (World Bank 2002). In a pioneering study in the Indian context, Jodha (1986) found that there is a negative relationship between the total income of the household and the share of total environmental income in it. This reflects that poorer households are more resource dependent.

Poor people are dependent more on ecosystem services of forest for their livelihoods. The common property resources such as forests generate income, food, medicine, tools, fuel, fodder, construction materials and so on. Poor people are liable to be severely affected when the environment is degraded. The forest dependent people depend on non-timber forest products like fuelwood, food, medicinal plants, masroom and honey etc., for their livelihood. Medicinal plants have an important role in rural health (Prasad and Bhatanagar 1993). In parts of West Bengal,

© The Author(s) 2017
J.P. Basu, *Climate Change Adaptation and Forest Dependent Communities*,
SpringerBriefs in Environmental Science, DOI 10.1007/978-3-319-52325-5_1

communities derive as much as 17 percent of their annual household income from Non-timber Forest Product (NTFP) collection and sale (Malhotra et al. 1991). The importance of NTFPs for the very poor tribal households has been well documented by other studies as well (Hedge et al. 1996; Godoy et al. 1995). In Andhra Pradesh, the poor obtain 84 percent of their fuel supplies from common property resources, and are employed for 139 days to collect products from common property resources (Jodha 1992). The potential NTFPs of the forests in South West Bengal are sal leaves, dry leaves, mushrooms, various medicinal plants, and kendu leaves etc. Among these green sal leaves are most important and many poor villagers depend upon this for their daily subsistence. It is found from the study conducted by (Mishra et al. 2006) that the average income from NTFPs in the region of Midnapore district of West Bengal is Rs. 550/- per household per year. There is large diversity of yearly income from NTFPs. Throughout India, collection of kendu leaf generates part time employment for 7.5 million people—a majority of them tribal women (Arnold 1995). According to studies in Uttar Pradesh, women derive a greater proportion of their income from forests and common lands; poor women derive 45% of their income from forests and common lands as opposed to 13% for men (FAO and SIDA 1991). The JFM program has an impact on enhancing income as well as livelihood and it is an instrument of rural development.

Non-Timber Forest Products (NTFPs) play an important subsistence and safety-net roles in the rural economy, but only a small subset of forest products possesses potential for significant cash income and employment generation (Wollenberg and Belcher 2001). A study by Wills and Lipsey (1999) in British Colombia estimated that in 1997 the commercial harvest of wild mushrooms, floral greens and other products employed almost 32,000 people on a seasonal or full-time basis, which generated direct business revenues of $ 280 million and overall provincial revenues in excess of $ 680 million. A study conducted by Grimes et al. (1994), showed that NTFP would contribute 77% to the annual net returns, if dry deciduous forests are exploited sustainably. The importance of NTFPs in the Hantana forest of Sri Lanka was INR 7052 per hectare per year for fuel wood collected from the forest while the value of grass was about INR. 578 (Abeygunawardena and Wikramasinghe 1992). A study of the Southern African Plateau in Botswana (Taylor and Parratt 1995) depicts that people most likely to be involved in NTFP use (namely rural communities) have very limited access to technology. Research by Sunderland et al. (1999) reconfirms that NTFPs provide sources of food, medicines, and income to many households in Central Africa. Pervez (2002) also observed that NTFPs in Dhading district of Nepal generated maximum employment (60.72%), followed by agriculture (22.30%), allied activities (15.83%) and other sources (1.16%).

Studies in India have revealed that NTFPs provide substantial inputs to the livelihoods of forest dependent communities many of whom have limited non agricultural income opportunities (Chandrasekharan 1998; FAO 1991). About 70% of the NTFP collection in India takes place in the tribal belt of the country (Mitchell et al. 2003). It would appear that the NTFP based small scale enterprises provide up to 50% of income for 20–30% of the rural labour force and 55% of employment in forestry is attributed to the sector alone (Joshi 2003). Rao and Singh (1996) studied the contribution of Non-wood forest products in augmenting the income of the tribal families in families of South Bihar and South West Bengal. Palit (1995) revealed that an average, each household of Raigarh forest protection committee was engaged for 63 days per year in the collection of NTFPs and earned INR. 2421 per household from the sale of NTFPs. India's economy is largely dependent on climate sensitive sectors such as agriculture, water resources and coastal zones, biodiversity and forestry (INCCA 2010). Further, findings of Intergovernmental Panel on Climate Change (IPCC 2001a, b) indicate that climate change is going to impose significant stress on resources. Specifically, with respect to India, preliminary assessments reveal that the severity of droughts and intensity of floods in various parts of India might increase (NATCOM 2004). Studies by Kumar et al. (2001) and Kumar Kavi (2002) point out that the inter-annual variability in rainfall will have major impact on food grain production in India and also on the economy of the country as a whole. Predicted increase in frequency and intensity of floods and droughts are likely to have unfavorable impacts on the occupational structure, food security, health, social infrastructure etc. of the hotspots (Roy et al. 2005). Kumar Nanda and Sutar's (2001) study showed that the forest dependent communities in Orissa were interested to protect forests and prevent forest fires because forests are of their sources of livelihoods.

1.2 Vulnerability

There is a considerable research work in the area of vulnerability to climate change (Adger 2006), much of it focused on conceptualizations of vulnerability and its relationship to adaptation. Many of the studies are simple conceptual diagrams useful for framing the issue of vulnerability to climate change at macro (global) scales (Füssel 2007a; Heltberg et al. 2009; Ionescu et al. 2009). Others focus explicitly on adaptation (Kelly and Adger 2000; Smit et al. 2000; Brooks 2003; Downing and Patwardhan 2004; Adger et al. 2005; Brooks et al. 2005; Füssel and Klein 2006; Füssel 2007b; Smit and Wandel 2005; Yamin et al. 2005).

The IPCC (2001a, b, c) defines vulnerability to climate change as 'the degree, to which a system is susceptible to, and unable to cope with, adverse effects of climate change, including climate variability and extremes. Vulnerability is a function of the character, magnitude, and rate of climate change and variation to which a system is exposed, its sensitivity, and its adaptive capacity' (Parry et al. 2007). Füssel (2007a, b, c) and Füssel and Klein (2006) argued that the IPCC (2001a, b, c) definition—which conceptualizes vulnerability to climate as a function of adaptive capacity, sensitivity, and exposure—accommodates the integrated approach to vulnerability analysis.

1.3   Conceptual Framework

There are three major conceptual approaches to analyze vulnerability to climate change. The first is socioeconomic, the second is biophysical (impact assessment) and the third is integrated assessment approaches.

1.4   Socioeconomic Approach

The socioeconomic vulnerability assessment approach mainly focuses on the socio-economic and political status of individuals or social groups (Adger 1999; Füssel 2007a, b, c). Individuals in a community differ in respects of education, gender, wealth, health status, access to credit, access and so on. In general, the socioeconomic approach focuses on identifying the adaptive capacity of individuals or communities based on their internal characteristics (Adger and Kelly (1999).

1.5   Biophysical Approach

The biophysical approach focuses on sensitivity (like change in yield, income and health etc.) to climate change. The impact of climate change on agriculture is measured by modeling the relationships between climatic variables and farm income (Mendelsohn et al. 1994; Polsky and Esterling 2001). Similarly, the yield impact of climate change is also analyzed by modelling the relationships between crop yields and climatic variables (Adams 1989; Kaiser et al. 1993; Olsen et al. 2000). Other related impact assessment studies are on food and water availability and ecosystem services (Du Toit et al. 2001; Xiao et al. 2002). This approach misses much of the adaptive capacity of individuals or social groups, which is more explained by their inherent or internal characteristics or by the architecture of entitlements, as suggested by Adger (1999).

1.6   The Integrated Assessment Approach

The integrated assessment approach combines both socioeconomic and biophysical approaches to determine vulnerability. The vulnerability mapping approach (O'Brien et al. 2004) in which both socioeconomic and biophysical factors are combined to indicate the level of vulnerability through mapping. In the IPCC framework, exposure has an external dimension, whereas both sensitivity and adaptive capacity have internal dimension, which is implicitly assumed in the integrated vulnerability assessment framework (Füssel 2007a, b, c).

1.7   Analytical Methods of Vulnerability

There are two types of analytical methods for measuring vulnerability. One is Indicator method and other is Econometric method.

Vulnerability studies based on Indicator method can be divided into two main branches depending on the use of weights (Cutter et al. 2000, Hahn et al. 2009, Kaly and Pratt 2000; Kaly et al. 1999, Easter 1999; Cutter et al. 2003, Deressa et al. 2008, Abson et al. 2012, Piya et al. 2012). Econometric method involves measuring the level of vulnerability using socio economic data sets from households. This method is divided into three categories like Vulnerability as uninsured exposure to risk (VER), Vulnerability as low expected utility (VEU) and Vulnerability as expected poverty (VEP). All of these three methods measure welfare loss due to shocks, but differ in that VEP and VEU measure ex-ante welfare loss whereas VER measures ex-post welfare loss due to shocks (Deressa et al. 2009).

The Environmental Vulnerability Index (EVI) developed by the South Pacific Applied Geoscience Commission (SOPAC) was one of the earliest efforts and examines vulnerability to environmental change (Kaly et al. 1999 and SOPAC 2005).

The Human Well-being Index (HWI) developed by Prescott-Allen (2001) is an alternative to the Human Development Index. The HWI incorporates demographic data as well as political rights, crime, internet users, and peace and order, and social equity gender and income.

The Prevalent Vulnerability Index (PVI) is a social vulnerability index that focuses on social, economic, institutional, and infrastructural capacity to recover from natural hazards or the lack thereof (Cardona 2005). In recent years, vulnerability assessment has become a noteworthy subject in the field of applied global change (McCarthy et al. 2001). The acknowledgement of a probable increase in the frequency and intensity of hazard events such as hurricane storm surge, flooding, and the potential exacerbation caused by sea level rise has yielded an increased interest in pre-hazard planning and emergency preparedness for climate related hazards (Wu et al. 2002; Adger et al. 2004; Rygel et al. 2006). Most of these studies focus on the physical dimensions of climate hazards (e.g. large scale exposure) (Adger et al. 2004; Brooks et al. 2005). Earlier assessments of the human dimensions of climate impacts focused more on specific impacts in developing countries, such as food scarcity (Bohle et al. 1994).

Acosta-Michlik et al. (2005) developed a framework for vulnerability assessment using the concept of security diagram. 'Security Diagram consists of three components, namely environmental stress, state susceptibility and crisis probability curves'.

Chakraborty et al. (2014) tried to categorize different regions of India according to their level of vulnerability. They identified districts in the state of Arunachal Pradesh, Assam, Bihar, Himachal Pradesh, Jharkhand, Manipur, Meghalaya, Mizoram, Uttar Pradesh, Uttarakhand, and West Bengal are the most vulnerable regions, while districts in the state of Punjab, Haryana, Gujarat, Tamil Nadu,

Maharashtra, Goa, Andhra Pradesh, Kerala, and Karnataka are among the least vulnerable regions.

Ravindranath et al. (2006) made an attempt to examine the impact of projected climate change on forest ecosystems in India. The study reached at the conclusion that under the climate projection for the year 2085, 77% and 68% of the forested grids in India are likely to experience shift in forest types under A2 and B2 scenarios respectively.

Chaturvedi et al. (2010) adopted Regional Climate Model of the Hadley Centre (HadRM3) and the dynamic global vegetation model for A2 and B2 scenarios to examine the impact of climate change on forestry of India. Result showed that 39% of forest grids are likely to undergo vegetation type change under the A2 scenario and 34% under the B2 scenario by the end of this century. Hazra et al. (2010) reported that forest area of Indian Sundarban has decreased from 2168.9 to 2132 km^2 during 2001–2008. The decrease in forest cover has serious impact of climate change and sea level rise.

Sarkar and Padaria (2010) attempted to identify farmers' awareness and risk perception about climate change in the coastal Indian Sundarban. The study revealed that people have more aware about increase in temperature, reduction in agricultural and livestock production, increase diseases and increase sea level etc.

Chaturvedi et al. (2011) projects the impact of climate change on Indian forests and conclude that 39% and 35% of the forests grids in India will likely undergo change under the A2 and B2 scenarios respectively.

O'Brien et al. (2004) estimated the vulnerability of India under multiple stressors (e.g. climate change, globalization etc.) and suggested North-western and central parts of India to be highly vulnerable and Southern India to be relatively low on vulnerability.

A study by Ravindranath et al. (2011) presents one of the best efforts to quantify vulnerability to climate change at the regional level. They assessed the district level vulnerability profiles of agriculture, water and forest sectors for North-eastern India and found that North Eastern India are vulnerable to climate change under the current climate and projected climate change.

1.8   Adaptation and Adaptive Capacity

Adaptations depend on adaptive capacity which represents the ways of reducing vulnerability. Adaptive capacities are assessed at a range of scales, from the individual (Marshall and Marshall 2007), household (Adger and Vincent 2005) and community levels of organization (Adger 2000; Berkes and Seixas 2006; Cinner et al. 2009c) to national assessments (Adger and Vincent 2005; Adger et al. 2005) Some approaches are inductive and use community-driven measures to assess capacity, whereas others are deductive and derived from the literature (Nelson et al. 2008). Some measures of adaptive capacity are best for comparing across scales (e.g. McClanahan and Cinner 2009), whereas others are more suitable for stand-alone assessments of specific communities or sectors. Several studies like Deressa et al. (2008) estimated the surrogated measures of exposure, sensitivity and adaptive capacity and then aggregated to generate an overall measure of vulnerability. Adaptive capacity is the ability of a system to adjust to climate change (IPCC 2001a, b, c). It describes the ability of a society or system 'to modify its characteristics or behavior so as to cope better with changes in external conditions' (Gbetibouo 2009).

Deressa et al. (2008) identified the determinants of adaptation using Heckman Probit selection model. Gbetibouo (2009) used Heckman Probit model and multinomial logit model to find out the determinant of adaptive capacity of 794 households in the Limpopo River Basin of South Africa for the farming season 2004–2005. The results of the multinomial logit and Heckman probit models revealed that household size, farming experience, wealth, access to credit, access to water, tenure rights, off-farm activities, and access to extension are the determinants that enhance adaptive capacity.

Amdu et al. (2013) identified local innovations for climate change adaptation and also find out determinants of adaptation collecting datasets from 384 households for 2010–2011 cropping season in Nile river basin. This study identifies adaptation options like implementation of soil conservation practices, cropping calendar adjustment, tree planting, change of crops, crop diversification and adjustment to crop and livestock management.

Ghosh (2012) identified 'migration' as the only adaptation options against climate change in Indian Sundarban region until 2012. Danda et al. (2010) pointed out establishment of knowledge centre, introduction of climate resilient agricultural and pisciculture practices, disaster relief shelter, helping early warning and response teams, partnership and networking are important adaptation strategy.

The level of interest in climate change adaptation can also be influenced by climate education and access to climate technology, expertise and information (Steinfeld 2001). An interest in adapting is necessary for individuals to identify the consequences, impacts and possible responses ("adaptation options") to climate change (Howden et al. 2007).

Resource users with dependents may be especially sensitive to climate changes and have a lower adaptive capacity since they will be less able to experiment with their options for the future and are consequently less flexible in their approach to change (Bennett 2001; Poggie and Gersuny 1974; Sorenson and Kaye 1999).

The income and debt levels of a resource-user and their ability to access credit can also significantly influence the extent to which a resource-user can effectively respond to change (Fisher 2001; Freudenberg and Frickel 1994; Johnson and Stallman 1994; Overdevest and Green 1995). Resource-users with a lower financial status often lack the flexibility with which to successfully absorb the costs of change and are often reluctant to take on further risks (Fisher 2001; Humphrey 1994; Nord 1994; Peluso et al. 1994).

Most of the above studies are useful for cross country comparison of vulnerability and fail to provide critical insights into effective adaptation strategies at the micro or household level. In addition, much of the early research work on adaptation focused on identifying potential impact of future climate change using General Circulation Models (GCMs). But the applications of such models are extremely limited in telling us about regional impacts of climate change and practical action on local level of adaptation.

Chapter 2
Data Base and Methodology

The data required for this purpose are varied which are collected from both primary and secondary sources. Different statistical techniques are applied to analyze the data. This chapter discusses the data base and methodology used for the analysis of the data.

2.1 Data Base

Data are collected from two sources for the study—Primary and secondary sources.

2.1.1 Primary Data and Methodology

This study was conducted in the three different agro-ecological regions of West Bengal namely, coastal Sunderban in the district of South 24 parganas, drought prone region in the district of Bankura and mountain region in the district of Darjeeling in West Bengal (Fig 2.1). The selection of villages in the coastal and drought prone areas was selected on the basis of socio-economic vulnerability. But the selection of villages in mountain area was purposely. One is village is in foothills and other is in the hill area.

The study was conducted in the two villages of Gossaba block in coastal Sunderban, West Bengal, namely Jamespur and Chargheri in 2011. The field work combined interviews and discussions with the local people and interviews with local experts and school teachers and other knowledgeable elders in the villages. The study selects 30% households randomly from each village. Total number of sample households in coastal Sunderbans was 202. Another study was conducted in the two other villages, namely Junsura and Baskula under Sonamukhi forest area in

© The Author(s) 2017   9
J.P. Basu, *Climate Change Adaptation and Forest Dependent Communities*,
SpringerBriefs in Environmental Science, DOI 10.1007/978-3-319-52325-5_2

Fig. 2.1 Position of different districts of West Bengal

the district of Bankura, one of the drought prone districts of West Bengal in 2011. In this area total sample household was 120. Similar study was conducted in the two villages namely, Khoirajhora forest basti which is situated on the foothills of Darjeeling and another is Rongtong (2) which is about1404 ft longitude in the mountain regions of Darjelling district in 2011. In this area total sample size was 71. A total of 393 structured households interviews for six villages taken together (Table 2.1).

Table 2.1 No. of sample households in different villages in different agro-climatic areas of West Bengal

Name of villages	No. of sample households	Agro-climatic areas of west Bengal
Junsura	60	Drought prone area
Baskula	60	Drought prone area
Jamespur	104	Coastal Indian Sunderbans
Chargheri	98	Coastal Indian Sunderbans
Khoirajhora Forest Basti	29	Mountain region of West Bengal
Rongtong (2)	42	Mountain region of West Bengal
Total	393	

Source: Field Survey 2011

2.2   Data Collection

Data on socio-economic variables, like age, sex, education, land holdings, sources of credit, physical assets, livestock assets, income from various sources, public health facilities, adaptation measures like migration, non-timber forest products; self-help groups have been collected from the field survey.

Construction of climate change Index: First, following the methodology of Cutter et al. (2000) and modifying the formula of Wu et al. (2002), the index value of each indicator of Climate change index is computed using the following formula.

$$In = \left(V_i \, / \, V_{max} \right).$$

Here, In is the index value of a particular indicator; V_i is the absolute value of a particular indicator; V_{max} is the maximum possible value of a particular indicator.

As no weight is attached to a specific indicator, Climate experience index is computed as the arithmetic mean of the respective (indicators) index values by giving the following formula.

$$I_X = \Sigma \, I_n \, / \, n$$

Here I_X is the climate change index; n is the number of indicators used to construct a particular index.

Climate change experience index/sea level rise index: The climate experience index has been formed consulting the paper of Hair et al. (2006). Villagers were asked different climate related questions and their answers were measured in a three point scale (3 = self realization, 2 = Heard from others, 1 = no idea). The acceptable limit of reliable coefficient is 0.70 (Hair et al. 2006).

2.2.1 Determinants of Adaptation by Heckman's Two-Step Model

We have used Heckman two- step model to estimate the determinants of adaptation to climate change. Adaptation to climate change is a two-stage process involving perception and adaptation stages. The first stage is whether the respondent perceived there was climate change or not, and the second stage is whether the respondent adapted to climate change on the condition that first stage that he/she had perceived climate change (Heckman 1976). To get information on their perceptions to climate change, people were asked question. The first was asking people if they have observed any change on the amount of temperature over the 10 years.

We have used maximum likelihood Heckman's two-step procedure (Heckman 1976) to correct for the selectivity bias.

Heckman's sample selection model assumes that there exists an underlying relationship which consists of:

The latent equation given by:

$$Yj^* = Xj\beta + U_{1j}. \tag{2.1}$$

Such that we observe only the binary outcome given by the probit model as:

$$Yj^{probit} = \left(Yj^* > 0\right). \tag{2.2}$$

The dependent variable is observed only if the observation j is observed in the selection equation:

$$Yj^{select} = \left(Zj\delta + U_{2j} > 0\right). \tag{2.3}$$

$$U_1 \sim N(0,1).$$

$$U_2 \sim N(0,1).$$

$$Corr\left(U_1, U_2\right) = \rho.$$

Where, x is a k- vector of explanatory variables which include different factors hypothesized to affect adaptation and z is an m vector of explanatory variables which include different factors hypothesized to affect perception; U_1 and U_2 are error terms. The first stage of the Heckman's sample selection model is the perceptions of climate change which is known as the selection model Eq. (2.3). The second stage is known as the outcome/adaptation model Eq. (2.1).

If we apply standard Probit techniques to Eq. (2.1), this will give rise to biased results. Thus, the Heckman probit provides consistent, asymptotically efficient estimates for all parameters in such models (Van de Ven and Van Praag 1981).

Chapter 3
Profile of Study Area and Socio-economic Analysis of Sample Households

After discussing the data and methodology in Chap. 2 it is necessary to describe the basic features of the study area and socio-economic conditions of sample households. Therefore, the present chapter is aimed at describing the different agro climatic regions of West Bengal say Drought prone region (the district of Bankura), the Coastal region of Indian Sunderbans (the district of South 24 Parganas) and Mountain region (the district of Darjeling) of West Bengal and the sample households.

3.1 About West Bengal

The State West Bengal is in the Eastern region of India. The geography of the state is unique in the sense that its northern part is in the Himalayan Range whereas the extreme southern part touches the Bay of Bengal and is covered by the Active Delta of the Sundarbans Mangrove forest. The greater part consists of alluvial plains.

The northern mountainous terrain covers Darjeeling and part of Jalpaiguri districts. The South-western peninsular part comprises Purulia, Medinipur, Bankura, Birbhum and part of Bardhaman districts.

The major rivers are Ganga flowing from west to east familiar with the name as Padma through Bangladesh and to south known as Bhagirathi. The other major rivers are Damodar-Kangsabati-Ajoy-Mayurrakshi etc. in the western part while Tista, Jaldhaka-Torsa-Raidak-Sankosh-Gangadhar are in the northern part of the state.

There are four seasons like (a) cold, dry weather from December to February; (b) hot, dry weather from March to May; (c) monsoon period from June to September; (d) post monsoon period in October and November. Over 70% of the rain falls between June and September.

The mean annual rainfall varies from 1026 mm. The state also has as long as 350 km of coastal line. The State of West Bengal is vulnerable to natural calamities like flood, cyclone, storm, thunder squall, drought, landslide, erosion and sometimes to earthquakes because of its geo-morphological, climatic and seismic conditions.

© The Author(s) 2017   13
J.P. Basu, *Climate Change Adaptation and Forest Dependent Communities*,
SpringerBriefs in Environmental Science, DOI 10.1007/978-3-319-52325-5_3

3.1.1 District Profile of the Drought Prone District of Bankura in West Bengal

Bankura, the fourth largest district of West Bengal is located in the western part of the state, is one the drought prone districts of West Bengal. The economy of Bankura district is dependent purely on agrarian. The district Bankura is agriculture-based economy and agriculture provides means of livelihood to 70% of population. The cropping pattern is towards paddy cultivation. There is a limited scope for the adaptation of mechanization due to un-conducive topography, very small size of the land holdings, poor irrigation coverage, low water retention capacity of soil etc. Fifteen years' (1995–2009) average actual rainfall is 1285 mm but normal rainfall is 1378 mm. Fifteen years' (1995–2009) average maximum temperature is 44.4 °C and minimum temperature is 8.2 °C. Agro-climatically, the region mainly occupies red and laterite soil zone. The trend of rainfall over 15 years is declining (see Fig. 3.1). The trend in maximum and minimum temperature for the district of Bankura is on the rise (see Figs. 3.2 and 3.3).

3.1.2 Socio-economic Conditions of Sample Households in Drought Prone Area

The socio economic characteristics of the sample households (in terms of caste, gender, age, education, male and female literacy rate, average family size, landholdings) are shown in Table 3.1. The socio-economic conditions are found to be very weak. From Table 3.1 it is clear that 20% households in the village of Junsura and 30% households in the village of Baskula have education up to primary level and only 3% in the village of Junsura and 8% in the village of Baskula have education up to secondary level. Seventy-six percentage of households in the village of Junsura

Fig. 3.1 Trends in rainfall in the District of Bankura

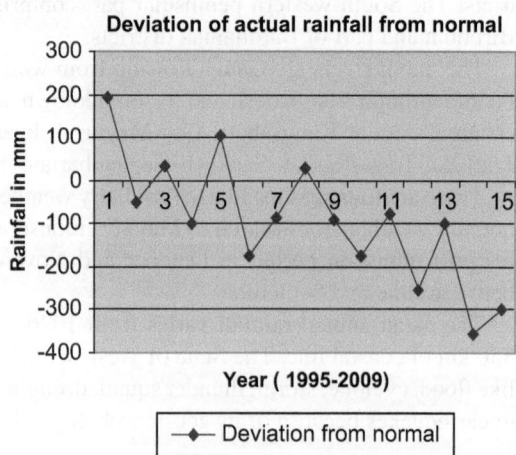

Deviation of actual rainfall from normal

Year (1995-2009)

Deviation from normal

Fig. 3.2 Trends in max
temperature

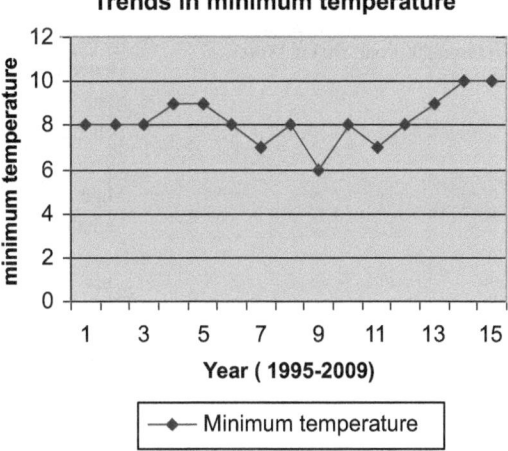

Fig. 3.3 Trends in
minimum temperature

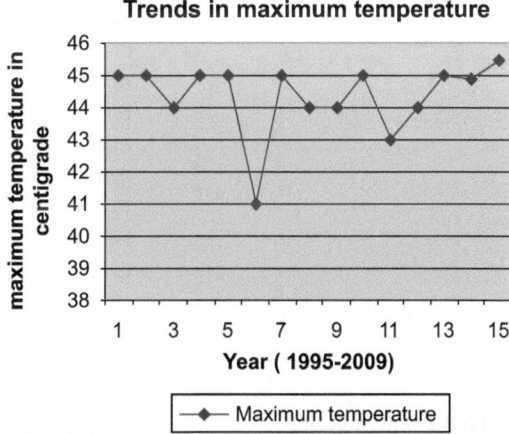

and 61% households in the village of Baskula are illiterate. Male literacy is 20% in the village of Junsura and 33% in the village of Baskula while female literacy rate lays 3–5% in both the villages. More than 80% households are found to be landless (Table 3.1). Only 15% households in the village Junsura and 5% households in the village Baskula have land holding up to 1 acre. Thus, the households in the villages are landless and marginal farmers.

The infrastructural availabilities such as electricity facility, facilities for drinking water, sanitation facilities, public health care facilities and borrowing facilities from banks or from moneylenders are presented in Table 3.2. This table shows that there are no public health care facilities, sanitation is provided by the local government and the households are under the private money lenders. Only 9% households in the village of Junsura and 1% households in the village of Baskula have borrowed loan from the bank.

Table 3.1 Socio-economic conditions of the Households in Drought prone area in West Bengal

	Junsura	Baskula
Caste		
General	–	–
Schedule caste	60(100)	–
Schedule tribe	–	60(100)
Gender		
Male	53(88.33)	49(81.67)
Female	7(11.67)	11(18.33)
Age		
≤30	16(26.67)	12(20.00)
31–40	17(28.33)	16(26.67)
41–50	13(21.67)	14(23.33)
51–60	9(15.00)	12(20.00)
Above60	5(8.33)	6(10.00)
Education		
Upto Primary (in 4-year)	12(20)	18(30)
Up to secondary (5–10 year)	2(3.3)	5(8.3)
Illiterate	46(76.7)	37(61.7)
Literacy rate		
Male	12(20)	20(33.3)
Female	2(3.3)	3(5)
Family size	3.9	2.9
Land holdings (in acres)		
Landless	51(85)	57(95)
≤1	9(15)	3(5)

Note: Figures in the parentheses represent percentage
Source: Field Survey 2011

Table 3.2 Infrastructural facilities in the drought prone regions of West Bengal

Drinking facility	Junsura	Baskula
Tube well (Yes)	60(100)	60(100)
No	–	–
Sanitation[a]facilities		
Yes	21(35.00)	25(41.67)
No	39(65.00)	35(58.33)
Public health care facilities		
Yes	–	–
No	60(100)	60(100)
Loan facilities		
From Banks	9(15.00)	1(1.67)
From money lenders	39(65.00)	57(95.00)
Not from any sources	12(20)	02(3.33)

Note: Figures in the parentheses represent percentage
[a]Sanitation facility given by the Panchayat
Source: Field Survey 2011

Table 3.3 Occupational structure of the households in the two villages in the drought prone regions

Name of the Villages			Household
Junsura	Single occupation	Only Forestry	3(5)
	Double occupation	Agriculture and forestry	1(1.7)
		Forestry and wage Labour	48(80)
	More than double occupation	Agriculture, forestry and wage labor	8(13.3)
	All		60(100)
Buskula	Single occupation	Only Forestry	4(6.7)
	Double occupation	Agriculture and forestry	1(1.7)
		Forestry and wage Labour	53(88.3)
	More than double occupation	Agriculture, forestry and wage labor	2(3.3)
	All		60(100)

Note: Figures in the parentheses represent percentage
Source: Field Survey 2011

Table 3.4   Preferences of livelihood of the households in the two villages

	Sourcesof livelihood	Agriculture	forestry	Wage labour
Junsura	First source	1(1.67)	55(91.67)	4(6.67)
	Second source	1(1.67)	5(8.33)	51(85.00)
	Third source	7(11.67)	0(0.00)	2(3.33)
Baskula	First source	0(0.00)	46(76.67)	14(23.33)
	Second source	2(3.3)	14(23.33)	38(63.33)
	Third source	1(1.7)	0(0.00)	2(3.33)

Note: Figures in the parentheses represent percentage
Source: Field Survey 2011

The occupational structure of the households is shown in Table 3.3. It is found from this table that 80% households in the village of Junsura and 88% in the village of Baskula are engaged in double occupation (either agriculture and forestry or forestry and wage labour). Thirteen percentage of households in Junsura and 3% households in Baskula are engaged in more than double occupation (agriculture, forestry and wage labour).

It is evident from Table 3.3 that agriculture, forestry and wage labour are the main sources of livelihood of the sample households in the drought prone regions of West Bengal. In addition households were asked to identify their sources of livelihood according to their own preferences. It has been found that 76–91% households have given preference on forestry (selling sal dish and fuel wood) as their first source of livelihood, 63–85% households wage labour as their second source of livelihood and only 11% household's agriculture as their third preference of livelihood in two villages (Table 3.4).

Table 3.5 Sources of income of the households in the two villages of drought prone areas

Sources of income	Junsura No. of households (n = 60)	Baskula No. of households (n = 60)
Agriculture	–	–
Sal dish	60(100)	58(96.7)
Fuel wood	38(63.33)	32(53.3)
Wage labour	58(96.7)	55(91.7)

Source: Field Survey 2011
Note: Figure in the parenthesis represents percentage of the total number of households

Table 3.6 Sources of Income of the households in the two villages of drought prone areas

	Land holding (acre)	Agriculture	Sal dish	Fuel wood	Wage income
Junsura	Landless	–	51(85)	34(56.7)	49(81.7)
	≤1	–	9(15)	4(6.6)	9(15)
Baskula	Landless	–	56(93.3)	29(48.3)	53(88.3)
	≤1	–	3(5)	3(5)	2(3.3)

Note: Figures in the parentheses represent percentage
Source: Field Survey 2011

Income is an important index which reflects economic status of the households. The sources of income by the households in the drought prone regions are shown in Table 3.5. It is observed from this Table 3.5 that all household (i.e. 100%) in the village of Junsura and 96% in Baskula earn income from selling sal dish*. Sixty-three percentage in the village of Junsura and 53% in the village of Baskula earn income from selling fuel wood and 96% in the village of Junsura and 91% in the village of Baskula earn income from wage labour.

It has been found from Table 3.6 that 85–93% landless households are making sal dish for income generation, 48–56% households collect fuel wood for income generation and 81% earn from wage labour in the both villages.

Mean annual income of the households in the two villages taken together is shown in the Table 3.7. The mean annual incomes of the households are Rs. 24,155 and Rs. 22,387 for the village Junsura and for the village of Baskula respectively. It is also revealed that income generation from sal dish activities is substantial to total income.

The asset distribution is shown in Table 3.8. Households have two types of assets like physical assets and livestock assets. Physical assets of the households comprise cycle, radio, agricultural input and mobile where as the livestock assets include cows, goats, hens and pigs. About 70% households in the village of Junsura and 73% households in the village of Baskula hold both physical and livestock assets.

Households mainly earn income from selling sal dish to local shops through the agents who collect the handmade sal dish and supply it to the market. Households collect the green sal leaves from the forests and stich the leaves with kutchi kathi which is also found in the forest. Then they dried the stiched leaves in the sunlight and give some weight on the dried leaves to make the dish round. This process is completed within one to 2 days. They generally get seven to eight rupees for the bunch which has eighty sal dishes. They can make average six bunches in one to 2 days interval and earn income twenty five rupees per day on the average.

Table 3.7 Mean annual
income of the households in
the two drought prone
villages (in Rs.)

Income	Junsura	Baskula
Mean annual income from agriculture	–	–
Mean annual income from sal dish	13,314	11,959
Mean annual income from fuel wood	6949	8691
Mean annual income from wage labour	6662	6537
Mean annual income	24,155	22,387

Source: Field Survey 2011

Table 3.8 Distribution of
Assets of the households in
the two villages

Assets	Junsura	Baskula
Only physical	–	1(1.7)
Only livestock	9(15)	12(20)
Physical and livestock	42(70)	44(73.3)
Land, livestock and physical	9(15)	3(5)
All	60(100)	60(100)

Note: Figures in the parentheses represent percentage
Source: Field Survey 2011

Table 3.9 livestock asset per household owned in the two villages

	Assets	No. of assets	No. of household	Asset per household owned
Junsura	Cows	71	39	1.82
	Goat	22	9	2.44
	Hen	202	47	4.30
	Pig	30	14	2.14
Baskula	Cows	91	41	2.22
	Goat	30	10	3.00
	Hen	264	51	5.18
	Pig	17	8	2.13

Source: Field Survey 2011

3.1.2.1   Distribution of Livestock assets

Distribution of livestock asset per household is shown in Table 3.9. In the village of
Junsura number of cows per household is 1 but it is 2 in the village of Baskula.
Number of goats per household is 2 in the village of Junsura and is 3 in the village
of Baskula. Number of hens per household is 4 in the village of Junsura and is 5 in
the village of Baskula which is the maximum among the livestock assets per household. A pig per household is 2 for both the villages (shown. in Table 3.9).

Table 3.10 Distribution of physical assets in two villages in the coastal Sunderbans

Physical asset	No. of households (n = 60) Junsura	No. of hhs (n = 60) Baskula
Cycle	47(78.33)	45(75.00)
Radio	8(13.33)	7(11.67)
Agricultural input	2(3.33)	5(8.33)
Mobile	6(10.00)	1(1.670)

Note: Figures in the parentheses represent percentage
Source: Field Survey 2011

3.1.2.2   Distribution of Physical Assets

Physical assets comprise cycle, radio, agricultural input (axe and spade) (used both for both in agriculture and household purpose) and mobile. Cycle is a means of transportation and almost available in each and every houses. It is seen from Table 3.10 that 78% households in Junsura and 75% households in Bakula own cycle. Eleven to thirteen percentage of households in the two villages have radio.

3.1.3   District Profile of the Coastal Indian Sunderbans District of South 24 Parganas in West Bengal

The Sundarbans constitutes an area of 26,000 km², of which 9630 km² is in Indian Territory and the rest in Bangladesh. The Indian component constitutes 106 islands, of which 54 are inhabited, located in 13 blocks in 24 Parganas South District and six blocks in 24 Parganas North District. According to the Census 2011 population of Indian Sundarban in the district of South 24 parganas is 3.3 million out of which male population is 1.7 million and female population is 1.6 million. 2.1 million out of 3.3 million people are found to be literate.

Indian Sundarbans is famous for mangrove forests. Sundarbans is associated with species diversity in terms of mangrove and mangrove related flora and fauna. The present forested part of Indian Sundarban region is controlled by Director of the Sundarban Biosphere Reserve (SBR) through the Divisional Forest Officer of South 24 Parganas Forest Divisions and the field Director of the Sundarban Tiger Reserve. The Sundarbans National Park within the Tiger Reserve was declared as World Heritage site in 1987 and is provided highest level of priority for protection. Figure 3.4 shows the forest cover area is lost every year in the Indian Sundarban area. There are 52 Forest Protection Committees and 14 Eco Development Committees in Indian Sunderbans.

Agriculture and fishing are two main sources of livelihood in the island areas of Sunderbans. About 95% of the population is primarily dependent on agriculture. About 50% of the farmers are landless and a large percentage of rural households

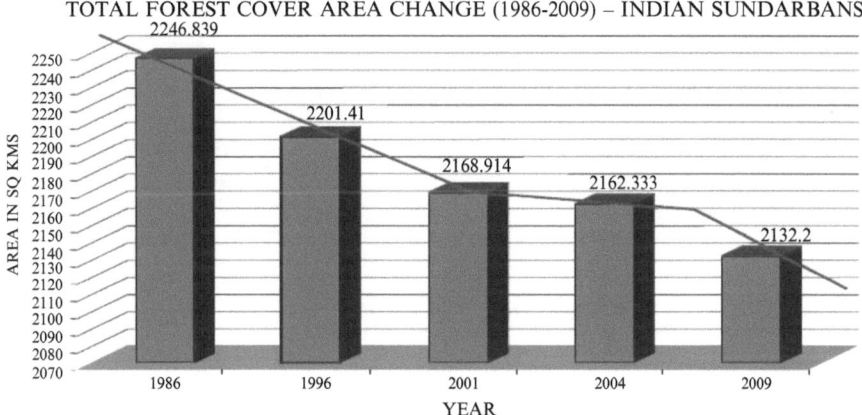

Fig. 3.4 Forest cover change in Indian Sunderbans.
Source: Danda et al. 2010

belong to marginal farmers. Many people of this region are dependent on the mangrove forest for their livelihood and sustenance through fishing, collection of honey and fuel wood. Another important source of income of the people is to collects seeds of tiger prawn throughout the year from Ganga Estuary system (Mahapatra et al. 1993). "A group of people locally known as 'Meendhara' is engaged with this job of collecting tiger prawn seeds. These people collect juvenile prawn and prawn seeds from brackish water. Presently 'Meen' collection has become economic mainstay of the villagers; however it only provides merely the level of subsistence" (Basu 2015).

The principal agricultural crops of the district South 24 parganas are Rice, Sunflower, Summer Moong, vegetables, fruits, coconut and betel nuts. A large number of people maintain their livelihoods from the agricultural sector (Comprehensive District Agriculture Plan under Rashtriya Krishi Vikas Yojana, South 24 Parganas District, West Bengal). Of 29 blocks more than 11 blocks of the district have saline or degraded alkaline soil. In major part of the district mono cropping pattern is practiced by the farmers due to the problem of water logging and poor irrigation facilities.

The climate of the Indian Sundarben region is mainly tropical. The average temperature in the district of South 24 parganas varies from a maximum of 38 °C to minimum of 13.5 °C. The annual average rainfall of the district is 1800 cm. The district receives more than 75% of the rainfall during monsoon. The trends in number of days with maximum temperature 35 °C or more is rising (shown in Fig. 3.5). Trends in annual (Fig. 3.6) total rainfall (1991–2009) is falling (shown in Table 3.6). Figure 3.7 shows the sea level rise has occurred in Sagar Island.

Fig. 3.5 Number of days
with maximum
temperature 35 °C or more
in South Twenty Four
Parganas.
Source: Basu 2015

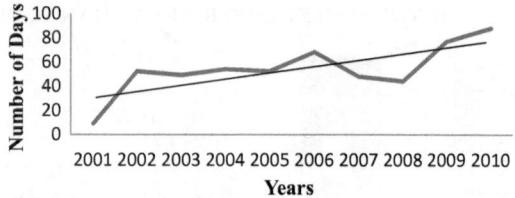

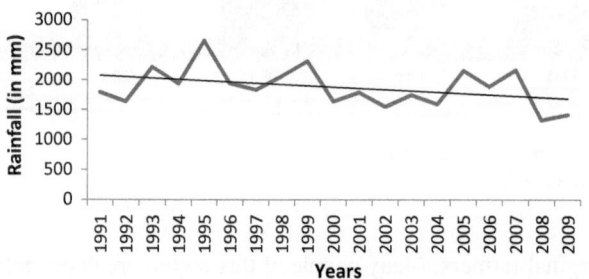

Fig. 3.6 Annual total rainfall in the district of South Twenty Four Parganas.
Source: Basu 2015

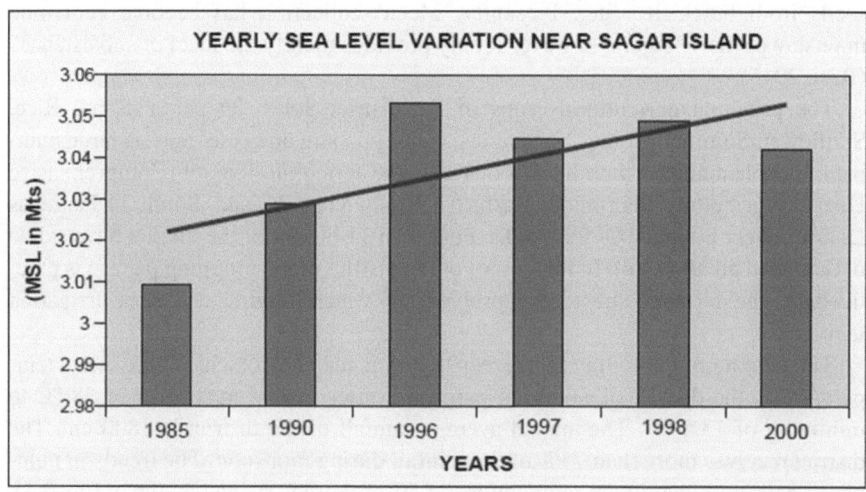

Fig. 3.7 Yearly Sea level variation in Sagar Island in the district of South Twenty Four Parganas.
Source: Danda et al. 2010

Table 3.11 Socioeconomic indicators in Coastal area of Sunderban

	Jamespur	Chargheri
Caste		
General	0(0)	5(5.1)
Schedule caste	104(100)	93(94.9)
Schedule tribe	–	–
Gender		
Male	98(94.2)	96(98)
Female	6(5.8)	2(2)
Age		
≤30	19(18.3)	13(13.3)
31–40	27(26.3)	38(38.8)
41–50	24(23)	24(24.5)
51–60	18(17)	11(11.2)
Above60	16(15.4)	12(12.2)
Education		
Upto Primary (in 4-year)	27(26)	19(19.3)
Upto secondary (5–10 year)	41(39.4)	48(49)
Higher secondary (10–12 year)	3(2.8)	1(1)
College (above 12 year)	0(0)	1(1)
Not through formal education	5(4.8)	2(2)
Illiterate	28(27)	27(27.7)
Literacy rate		
Male	75(72.1)	70(71.4)
Female	1(0.96)	1(1)
Family size	4.4	4.3
Male	175	146
Female	152	140
Children	131	132
Land holdings (in acres)		
Landless	58(56)	77(78.6)
≤1	42(40.2)	19(19.4)
1–2	3(2.9)	2(2)
Above 2	1(0.9)	–

Note: Figures in the parentheses represent percentage
Source: Field Survey

3.1.4   Socio-economic Profile of Sample Households in Coastal Area

Socio-economic indicators of the sample households in the two coastal villages in
Sunderbans are presented in Table 3.11. The socio-economic conditions are very
poor in the coastal Sunderbans. The village Jamespur is schedule caste dominated
village (100% households are schedule caste) while Chargheri is both schedule

Table 3.12 Infrastructural facility in the two coastal villages of Sunderbans

Drinking facility	Jamespur	Chargheri
Tap	104(100)	98(100)
Well	–	–
*Sanitation*facilities*		
Yes	104(100)	98(100)
No	0(0)	0(0)
Public health care facilities		
Yes	–	–
No	104(100)	98(100)
Loan facilities		
From Banks	12(11.5)	16(16.3)
From money lenders	91(87.6)	81(82.7)
Not from any sources	1(0.9)	1(1)
Solar light facility		
Yes	1(1)	0(0)
No	103(99)	98(100)

Note: Figures in the parentheses represent percentage
Source: Field Survey
*Sanitation facility given by the Panchayat

caste and general caste dominated (94% Schedule caste, 5% General caste respectively) village. More than 39% households completed their education up to secondary level. It is revealed from Table 3.11 that more than 70% male population is literate in both the villages. More than 50% households are landless in the village Jamespur while more than 75% households are landless in the village Chargheri. In both the villages there is predominance of marginal farmers (Table 3.11).

The infrastructural accessibility of the sample households in the two villages is shown in Table 3.12. In our study it has been found that all the households have sanitation facility .This facility has been provided by the Tagore Society, an NGO. There is no public health care service in the two villages. Households usually borrow loan from money lenders at a higher interest rate (87% in the village of Jamespur and 82% in the village of Chargheri).A few number of households use banking facilities (11% in the village of Jamespur and 16% in the village of Chargheri). In both villages do not have electricity. Only 1% households in the village of Jamespur use solar light as an alternative use of electricity.

Households have been categorized as single occupation households (who are generally engaged in either only agriculture or only wage labour or only fishing and crabs), double occupation households (who are engaged in either agriculture and fishing or Agriculture and wage labour or fishing and wage labour simultaneously) and more than double occupation households (who are engaged in agriculture, fishing and wage labour). Larger percentage of households is engaged in more than double occupation (39% households in the village of Jamspur and 72% households in the village of Chargheri) (Table 3.13). This shows the diversification of occupation prevailing in the two villages.

Table 3.13 Occupational structure of the households in the two villages of coastal areas

Name of the villages			Household
Jamespur	Single occupation	Only agriculture	1(0.9)
		Only wage labor	4(3.8)
		Only fishing and crab collection	12(11.5)
	Double occupation	Agriculture, fishing and crab collection	13(12.5)
		Agriculture and wage labour	6(5.8)
		Fishing and crab collection and wage labour	41(39.5)
	More than double occupation	Agriculture, fishing and crab collection and wage labor	23(22.2)
		Agriculture, fishing andhoney collection	4(3.8)
	All		104(100)
Chagheri	Single occupation	Only agriculture	–
		Only wage labor	3(3)
		Only fishing and crab collection	1(1)
	Double occupation	Agriculture and fishing and crab collection	–
		Agriculture and wage labour	5(5.1)
		Fishing and crab collection and wage labour	71(72.5)
	More than double occupation	Agriculture, fishing and crab collection and wage labor	18(18.4)
		Agriculture, fishing and crab collection and honey collection	–
	All		98(100)

Note: Figures in the parentheses represent percentage
Source: Field Survey

The sources of livelihood as revealed from Table 3.13 are agriculture, fishing and crab collection, honey collection and wage labour. Households were asked to identify their sources of livelihood and were called upon to arrange livelihoods in order of their preferences. It has been found that the households' first preference is fishing and crabs collection (72% in the village of Jamespur and 60% in the village of Chargheri) (Table 3.14).The second source of livelihood is wage labour (35% in the village of Jamespur and 55% in the village of Chargheri) and agriculture is their third source of livelihoods (14% in the village of Jamespur and 10% in the village of Chargheri) (Table 3.14).

Table 3.14 Preferences of livelihood of the households in the two coastal villages

	Sources of livelihood	Agriculture	Wage labour	Honey collection	Fishing and crabs	Van puller
Jamespur	First source	3(2.9)	29(27.9)	0(0)	75(72.2)	1(.95)
	Second source	24(23.1)	37(35.6)	9(7.7)	12(11.5)	2(1.9)
	Third source	15(14.4)	8(7.7)	1(0.95)	2(1.9)	0(0)
Chargheri	First source	2(2.04)	36(36.7)	1(1)	59(60.2)	0(00)
	Second source	10(10.2)	54(55.1)	1(1)	29(29.6)	0(0)
	Third source	10(10.2)	5(5.1)	0(0)	1(10)	0(0)

Note: Figures in the parentheses represent percentage
Source: Field Survey

Table 3.15 Distribution of Sources of income of the sample households in the two coastal villages

	Jamespur	Chargheri
Sources of income	No. of households (n = 104)	No. of households (n = 98)
Agriculture	4(3.84)	2(2)
Wage labour	74(71)	95(97)
Honey collection	11(10.6)	2(2)
Fishing and crab collection	90(86.5)	89(90.8)

Note: Figures in the parentheses represent percentage
Source: Field Survey

3.1.4.1 Description of Different Components of Income

Agriculture is produced once a year. A little amount is left for selling after meeting self consumption. Wage labour is the important source of income. But the households are not getting income for the entire year because of the unavailability of jobs in the study areas. They are mainly agricultural labour. They also reported that they get 6 months working days on the average in a year and 20 days per month. They get wage of Rs 100/- per day. Honey is collected from the forest once a year. They collect the permit first from forest department and a group of 5–6 people enter into the forest. Honey collection continues for 12–15 days in a year. They collect 1–1.5 quintal of honey per visit. Per head collection is about 20 kg. They have to face many risks like attack of tiger and other wild animals in the forest while they are going to honey collection. They are liable to sell the honey to forest department at a lower cost at the rate of Rs 40/- per kilogram of honey while market price was Rs 100–120 per kilogram of honey.

Fish and crabs also are collected from the mangrove forest. They reported that collection of fish and crabs continues for 8 months a year. They generally go to the forest by forming a group of 3 men. The total collection of fish and crabs is about 60–70 kg per month. Their per head collection of fish and crabs are nearly 20 kg in each month. The price of fish and crabs is Rs 46/- per kilogram.

The sources of income by households are presented in Table 3.15. It is found that 86% households in the village of Jamespur and 90% households in the village of

Table 3.16 Distribution of different sources of income by land holdings

	Land holdings (in acre)	Agriculture	Wage labours	Honey collection	Fishing and crabs
Jamespur	Landless	–	41(39.42)	5(4.81)	53(50.96)
	≤1	42(40.38)	25(24.04)	7(6.73)	33(31.73)
	1–2	3(2.88)	3(2.88)	–	2(1.92)
	Above 2	1(0.96)	–	–	1(0.96)
Chargheri	landless	–	74(75.51)	2(2.04)	73(74.49)
	≤1	19(19.39)	18(18.37)	–	14(14.29)
	1–2	2(2.04)	2(2.04)	–	1(1.02)
	Above 2	–	–	–	–

Note: Figures in the parentheses represent percentage
Source: Field Survey

Table 3.17 Mean annual income of the households in Coastal areas of Sunderbans

Income	Jamespur	Chargheri
Mean annual income from agriculture	2000	2810
Mean annual income from wage labour	883.78	7663.6
Mean annual income from honey collection	868.2	2750
Mean annual income from fishing and crabs	12,360	7762.7
Total Mean annual income	17,961	14,592

Source: Field Survey

Chargheri earned income from selling fish and crabs. Seventy-one percentage of households in the village of Jamespur and 97% households in the village of Chargheri earned income from wage labour. About 11% households are engaged for honey collection in the village Jamespur while 2% households in the village of Chargheri (Table 3.15).

From Table 3.16 it is found that 39% landless households in the village of Jamespur and 75% landless households in the village of Chargheri earned income from selling wage labour. Fifty percentage of landless households in the village of Jamespur and 74% landless households in the village of Chargheri earned income from selling fish and crabs (Table 3.16).

Mean annual income is Rs.17961.00 and Rs.14592.00 for the villages Jamespur and Chargheri respectively shown in Table 3.17. The mean annual income from fishing and crabs is substantially higher in both the villages.

Livestock ownership is considered to be an activity that engages households to earn income from home. Households thus contribute significant labour for livestock, around 2–5 h daily. They generally rear cows, goats, hens, pigs and sheep which help them to be economically independent. In our study it is seen that households own more than 3 hens and duck for the two villages (shown in Table 3.18).

Table 3.18 livestock asset per household own in the two villages of coastal sunderbans

	Assets	No. of assets	No. of Household	Asset per household
Jamespur	Cows	57	72	0.8
	Goat	86	63	1.4
	Hen	209	55	3.8
	Duck	71	20	3.5
	Pig	3	2	1.5
	Sheep	10	6	1.7
Chargheri	Cows	34	17	2
	Goat	59	22	2.7
	Hen	149	51	3
	Duck	106	28	3.8
	Pig	0	0	0
	Sheep	2	1	2

Source: Field Survey

Table 3.19 Distribution of physical assets in the two coastal villages Sunderbans

Physical asset	No. of households (n = 104) Jamespur	No. of households (n = 98) Chargheri
Fishing net	64(61.5)	78(79.6)
Boat	38(36.5)	44(44.5)
Cycle	15(14.4)	6(6)
Van	3(2.9)	0(0)
Agricultural input (axe, spade etc.)	19(18.3)	26(26.5)
Radio	14(13.7)	15(15.3)
Mobile	12(11.8)	4(4)

Note: Figures in the parentheses represent percentage
Source: Field Survey

Physical assets consist of fishing net, motor boat, bicycle, agricultural inputs, van, radio and mobile. Almost all households have fishing net (61% households reported in the village of Jamespur and 79% households reported in the village of Chargheri). Fishing net is generally used by the male members. Most of the households don't use motor boat because its cost is too high. Only 36% households in the village Jamespur and 44% households in the village Chargheri used manual boat. Cycle is very rarely used by the villagers (14% in the village of Jamespur and 6% in the village of Chargheri used cycle).Their only means of transportation is boat. About 13% in the village of Jamespur and 15% in the village of Chargheri have physical assets like radio for getting information on weather (Table 3.19).

3.1.5   District Profile of the Mountain District of Darjeeling in West Bengal

Darjeeling Himalaya is a part of eastern Himalayan ranges and is bounded by Sikkim, Nepal and Bhutan on the North, West and East respectively. The area covered by Darjeeling Himalaya is about 1721 Km^2. while the total area of the district is about 3202 Km^2. The annual mean maximum temperature and minimum temperature lies between 12 and 1.7 °C and the average annual precipitation in Darjeeling is about 3000 mm.

Darjeeling Himalaya is constituted by over 13% cultivable land in proportion to its total geographical area. Agricultural crops in the Himalaya are classified into two categories like food crops and cash crops. Rice, maize, potato, wheat, barley etc. are food crops while the cash crops are tea, cinchona, ginger etc. Fruits such as orange, papaya, peaches, guava, plumbs and even mangoes are grown in the valleys and in areas with low altitudes. Livestock and animal husbandry engages a significant proportion of rural folks in the area.

Darjeeling is famous for its tea production and exporting it in the world market. Tea, tourism and timber are the backbone of the hill economy. Women constitute the bulk of the labour force in tea gardens, given their skills in plucking the leaves form the tea bush (Taknet 2002).

The Darjeeling hill is highly landslide risky area. Heavy monsoon precipitation is a very common cause of the landslides. About 30% of the forest covers found in the lower hills are deciduous. Evergreen forest constitutes only about 6% of the total forest coverage. A study was done by the "Save the hills", an organization working on disaster prevention relating to landslide in Darjeeling recorded the data from 1889 to 2009. The report of landslides is presented Table 3.20.

Rainfall pattern in the district of Darjelling last 5 years is shown in Fig. 3.8. The trend is found to be increasing.

3.1.6   Socio-economic Profile of Sample Households in Mountain Area

The socio-economic indicators of the sample households in the two villages in the district of Darjeeling are shown in Table 3.21. This table provides the information that the both villages are dominated by other backward classes, and scheduled tribes' people. In the village of Khoirajhora forest basti 37% households are schedule tribes while in Rongtong (2) the percentage is 54(Table 3.21).

In the village of Khoirajhora forest basti the maximum households attained up to primary level of education (41.5%) whereas in the village of Rongtong (2) the larger proportion of households attained up to secondary level of education (40.4%).About 17.2% and 38.2% households are illiterate in Khoirajhora forest basti and Rongtong (2) respectively (Table 3.21). Male literacy rate is higher in khoirajhora forest basti

Table 3.20 Landslide occurrences in Darjeeling Himalayas

Period	Nature of damage/casualties	Area affected
Sep1899	72 persons died	Darjeeling town
Jan1934	Not available	Darjeeling district during the Bihar-Nepal earthquake
Jun1950	Kalimpong component of Darjeeling Hill Railways closed permanently	Entire district
Oct1968	Thousands died in Darjeeling district and Sikkim	Incessant rains for 4 days triggered 20,000 landslides in Darjeeling and Sikkim
Jul1993	Not available	Affected Rangli Rangliot tea estate
Jul1996	36 persons died	Affected Nimbong (Kalimpong subdivision)
Aug1997	12 died, water supply disrupted	Darjeeling town
July 1999	Not available	Kurseong town
July 2003	17 people died, many houses were lost	Mirik
Sep2007	5 died in Kalimpong, 12,000 people affected, 1150 houses	Darjeeling district and Sikkim
May2009	35 deaths, 4500 houses fully damages, 12,000 houses partly damaged	Entire district

Source: A study done by Save the Hills, an organization working on disaster prevention relating to landslides in the Darjeeling hills.

Fig. 3.8 Rainfall in the district of Darjeeling

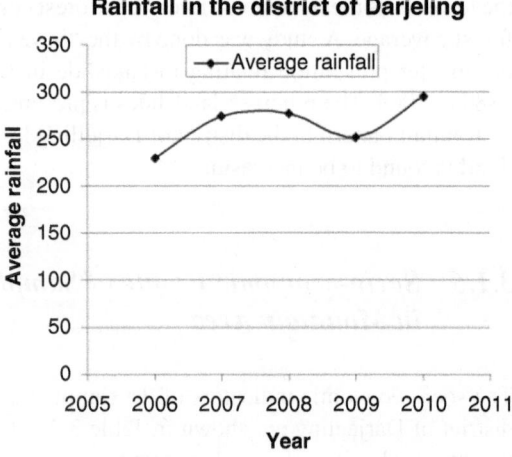

Rainfall in the district of Darjeling

than that in Rongtong (2). Family size is found to be 4.45 and 4.2 for the villages of Khoirajhora forest basti and Rongtong (2) respectively. About 51% households in the village of Khoirajhora forest basti and only 21% households in the village of Rongtong (2) accounted for holding land upto 1 acre. Most of the households i.e., 78.5% are landless households in Rongtong (2).

Table 3.21 Socio-economic indicators in Mountain region of West Bengal

Caste	Khoirajhora forest basti	Rongtong (2)
General	2(6.8)	7(16.6)
Schedule caste	1(3.4)	4(9.5)
Schedule tribe	11(37.9)	23(54.8)
Other backward class	15(51.7)	8(19.1)
All	29(100)	42(100)
Gender		
Male	24(82.7)	25(69.5)
Female	5(17.3)	17(30.5)
All	29(100)	42(100)
Age		
≤30	9(31.1)	10(23.8)
31–40	8(27.5)	7(16.6)
41–50	7(24.1)	16(38.1)
51–60	3(10.5)	5(11.9)
Above60	2(6.8)	4(9.6)
All	29(100)	42(100)
Education		
Upto Primary (in 4-year)	12(41.5)	5(11.9)
Upto secondary (5–10 year)	10(34.5)	17(40.4)
Higher secondary (above 10 year)	2(6.8)	4(9.5)
Illiterate	5(17.2)	16(38.2)
All	29(100)	42(100)
Literacy rate		
Male	20(68.9)	20(47.6)
Female	4(13.7)	6(14.2)
Family size	4.45	4.2
Land holdings (in acres)		
Landless	8(27.5)	33(78.5)
≤1	15(51.7)	9(21.5)
2–5	5(17.2)	0
Above 5	1(3.6)	0

Note: Figures in the parentheses represent percentage
Source: Field Survey 2011

3.1.6.1 Infrastructural Facilities

The accessibilities of infrastructure are presented in Table 3.22. Drinking water is scare in the mountain areas. Public Health Engineering (PHE) supplies the water through pipe line and the people of the two villages use it (100% households reported in both the villages). But there is a shortage of water supply from July to September and sometimes it is totally interrupted when landslides or other natural disasters occur. At the time of landslides people of the two villages bring water from lake or tube well using big drum on their back. It is comparatively difficult for the

Table 3.22 Infrastructural facilities in mountain regions

Infrastructural facility	Khoirajhora forest basti	Rongtong (2)
Drinking facility		
Tube well/P.H.E. supply (Yes)	29(100)	42(100)
No	0	0
All	29(100)	42(100)
Sanitation facilities		
Yes	27(93.1)	41(97.6)
No	2(6.9)	1(2.4)
All	29(100)	42(100)
Public health care facilities		
Yes	0	0
No	29(100)	42(100)
All	29(100)	42(100)
Loan facilities		
From Banks	0	0
From money lenders	15(51.7)	19(45.2)
Not from any sources	14(48.3)	23(54.8)
All	29(100)	42(100)

Note: Figures in the parentheses represent percentage
Source: Field Survey 2011

villagers of Rongtong (2) as it situates on 1404 ft altitudes from the plain. For this reason the villagers of Rongtong (2) harvest rain water to manage their future water-shortage. About 93% households in the village of Khoirajhora forest basti and 97% in the village of Rongtong (2) have sanitary facility at home (Table 3.22).

Health care service is an essential need of the people. But in the two studied villages it is found that medical facilities are rare (Table 3.22). Villagers reported that they generally have to travel 8–10 km for medical facilities.

People of the two villages are not getting the advantage of banking services. For this reason, people borrow money from money lenders (51.7% in the village of Khoirajhora forest basti and 45% in the village of Rongtong (2)).

Land is a primary source of occupation. The villagers work hard to make the hill area suitable for cultivation. They generally cut the hill and make it plain for their self consumption. The people of the two villages cultivate mainly rice, maize, zinger and *kucho*.Rice is produced in the khoirasjhora forest basti but the villagers of Rongtong (2) are unable to produce rice due to natural condition and unavailable source of water. They mainly produce zinger, maize and kucho. The local name Kucho is used for the purpose of home cleaning.

Subsistence agriculture, animal husbandry, wage labour, and services are the major activity of the rural people. In the village of Khoirajhora forest basti households generally earn income from double sources (animal husbandry and wage labour) and more than double occupation (agriculture, animal husbandry and wage labour). In the village of Rongtong (2) households have double occupation (45%

Table 3.23 Occupational structure of the households in the two villages

	Occupation	Khoirajhora forest basti	Rongtong (2)
Single occupation	Only agriculture	0	0
	Only animal husbandry	0	0
	Only wage labour	0	7(16.6)
	Only Service	0	2(4.7)
Double occupation	Agriculture and animal husbandry	0	3(7.1)
	Agriculture and wage labour	1(3.4)	4(9.5)
	Agriculture and service	0	1(2.3)
	Animal husbandry and wage labour	7(24.1)	19(45.2)
	Animal husbandry and service	0	4(9.5)
	Wage labour and service	1(3.4)	1(2.3)
More than double occupation	Agriculture, animal husbandry and wage labour	20(69.1)	1(2.8)

Note: Figures in the parentheses represent percentage
Source: Field Survey 2011

Table 3.24 Preferences of livelihood of the households in the two villages

	Sources of livelihood	Agriculture	Animal husbandry	Wage labour	Service
Khoirajhora forest basti	First source	8(27.5)	1(3.4)	18(62.1)	2(6.8)
	Second source	13(44.8)	7(24.1)	7(24.1)	1(3.4)
	Third source	0	19(65.5)	1(3.4)	0
	no source	8(27.7)	2(7)	3(10.4)	26(89.8)
	All	29(100)	29(100)	29(100)	29(100)
Rongtong (2)	First source	4(9.5)	2(4.7)	29(69.1)	8(19.1)
	Second source	5(11.9)	23(54.7)	4(9.5)	0
	Third source	0	6(14.2)	0	0
	no source	33(78.6)	11(26.4)	9(21.4)	34(80.9)
	All	42(100)	42(100)	42(100)	42(100)

Note: Figures in the parentheses represent percentage
Source: Field Survey 2011

households in animal husbandary and wage labour). Only 4.7% households of
Rongtong (2) are engaged in the service sector (Table 3.23).

There are four sources of livelihood of the sample households like agriculture,
animal husbandry, wage labour and services (Table 3.24). Households were asked
to rank their sources of livelihood as per their preferences. About 62% households
in the village of Khoirajhora forest basti and 69% households in the village of
Rongtong (2) reported that they earn income mainly from wage labour (Table 3.24).
They get jobs for tree planting, cleaning of forest, road construction and repairing;
work in the tea garden etc. They do not have job opportunities for entire year.

Table 3.25 Distribution of sources of income in mountain regions of Darjeeling

Sources of income	Khoirajhora forest basti No. of households (n = 29)	Rongtong (2) No.of households (n = 42)
Agriculture	4(13.7)	8(19.1)
Animal husbandary	21(72.4)	27(64.3)
Wage labour	26(89.6)	32(76.1)
Service	3(10.3)	8(19.1)

Source: Field Survey 2011
Note: Figure in the parenthesis represents percentage of the total number of households

Table 3.26 Mean annual income (in Rs.) of the households in mountain regions of Darjeeling

Income	Khoirajhora forest basti	Rongtong (2)
Mean annual income from agriculture	4950	2156.25
Mean annual income from animal husbandry	2728	2140.7
Mean annual income from wage labour	10940.8	10766.3
Mean annual income from service	44,000	72,450
Total mean annual income	17,019	23,790

Source: Field Survey 2011

Table 3.27 Livestock asset per household in the two villages of mountain areas

Khoirajhora forest basti Assets	No. of asset	No. of hhs	Asset per hhs
Cows	48	18	2.67
Goat	13	8	1.63
Hen	117	21	5.57
Rongtong (2)			
Cows	11	5	2.20
Goat	58	18	3.22
Hen	229	31	7.39

Source: Field Survey 2011

Villagers' second and third sources of livelihood are agriculture and animal husbandry respectively in the two villages.

Animal husbandry and wage labour are the main sources of income in both the villages. About 72.4% households in the village of Khoirajhora forest basti and 64.3% households in Rongtong (2) are deriving income from animal husbandry while 89.6% households in Khoirajhora forest basti and 76.1% households in Rongtong (2) earn income from wage labour (Table 3.25).

In Table 3.26 we have estimated the mean annual income from different sources. Mean annual income is Rs17019 for the village of Khoirajhora forest basti while Rs 23,790 for the village Rongtong (2) (Table 3.26).

In our study it is seen that households own more than 5 hens and duck for the two villages (shown in Table 3.27). Households in the two villages also own more than 2 cows. These livestock assets help to support financially to improve their living conditions.

Table 3.28 Distribution of physical assets of the households in the two villages

	Khoirajhora forest basti	Rongtong (2)
Physical asset	hh(29)	hhs(42)
Cycle	25(86)	1(2.3)
Van	2(6.8)	0
Radio	2(6.8)	5(11.9)
Agricultural input	12(41.3)	1(2.3)
T.V	16(55.1)	30(71)
Mobile	15(51.7)	28(66.6)

Note: Figures in the parentheses represent percentage of total number of households
Source: Field Survey 2011

Households own different types of physical assets like cycle, van, radio, agricultural input, Television (TV) and mobile for their usage. Cycle is a basic mean of transport for the households of Khoirajhora forest basti (86% households have cycle) where as only 2% households have cycle in Rongtong (2). Holding of T.V and mobile are more in Rongtong (2) than in the village of Khoirajhora forest basti (Table 3.28).

Chapter 4
Vulnerability Analysis

The aim of this chapter is to present key vulnerability assessments along with measurement of socio-economic vulnerability. There are two main approaches to vulnerability assessments. One is impact-based approach and other is vulnerability-based. This chapter discusses first generation and second generation vulnerability assessments along with the study of vulnerability by security diagram and Fuzzy Inference system. In addition this paper attempts to measure the socio-economic vulnerability for drought, coastal and mountain regions of West Bengal.

4.1 First Generation and Second Generation Vulnerability Assessments

There are two ways of assessing vulnerability of climate change. First is hazards-based adaptation approach which is associated with the response to the observed and experienced impacts of climate change on society. The second is vulnerability reduction—based adaptation approach which is called as second generation vulnerability assessment. In the hazards-based adaptation approach, responses ensure that the vulnerability to the impact is reduced. This in turn leads to the reduction of risk. With reduced risk development takes place in more sustainable ways. In short, the process is given below:

Adaptation *to* climate change impacts → Vulnerability reduction → Development

Vulnerability-based approach which is known as "second-generation vulnerability assessments" emphasizes various non climatic factors of vulnerability and adaptive capacity which includes poverty, economic inequality, effectiveness

J.P. Basu, *Climate Change Adaptation and Forest Dependent Communities*,
SpringerBriefs in Environmental Science, DOI 10.1007/978-3-319-52325-5_4

of government institutions, literacy, and education levels. The primary advantage of this approach is that it incorporates both climatic and non climatic vulnerability factors into adaptation planning. In this approach priority is given first development which can reduce vulnerability to climate change. By reducing the vulnerability, impacts of climate hazards can be minimized. This further transforms into a process of adaptation to climate change. In short, the process is given below:

Climate – aware development→ Vulnerability reduction → Impact reduction → Adaptation

Vulnerability assessment is an extension of a climate change impact assessment. This assessment is classified into two headings, viz. first generation vulnerability and second generation vulnerability. Figure 4.1 shows the main components of the first generation vulnerability assessment.

Climate variability is a new component in Fig. 4.1. Global climate change affects climate variability in terms of frequency, intensity and location of extreme events. In addition, non-climatic factors like economic, social, demographic, technological and political factors affect both sensitivity and exposure to climate change stimuli (Fig. 4.1).

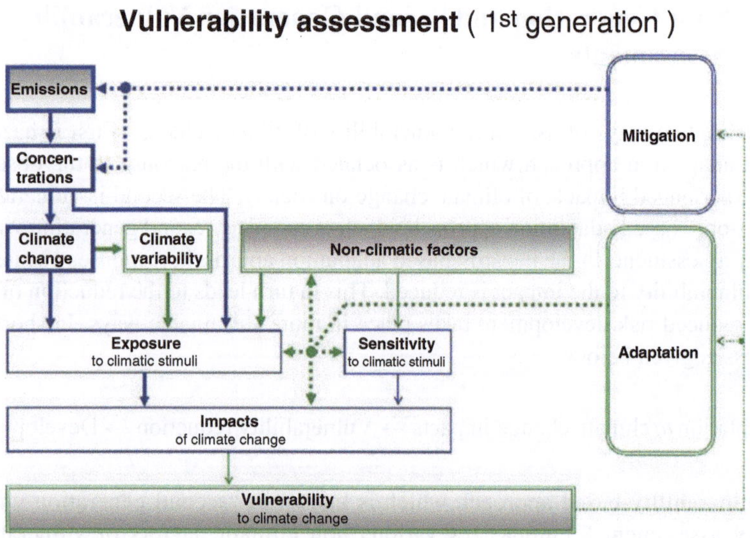

Fig. 4.1 Conceptual frameworks for a first generation vulnerability assessment.
Source: Füssel and Klein 2006

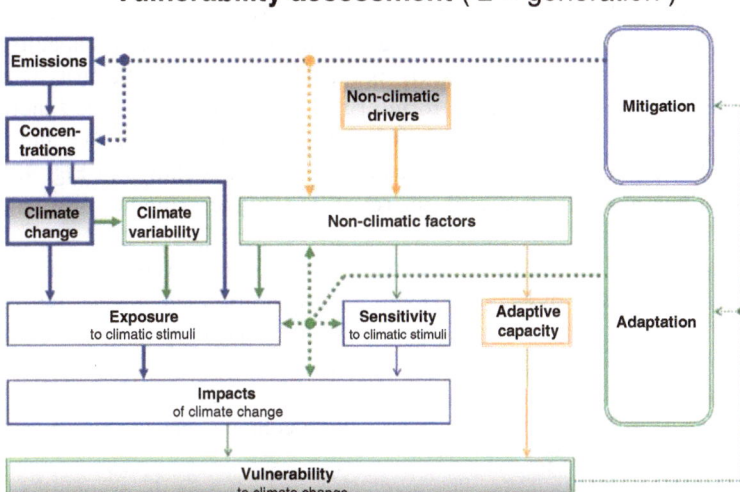

Fig. 4.2 conceptual frameworks for a second generation vulnerability assessment.
Source: Füssel and Klein 2006

The following Fig. 4.2 presents the components of second generation vulnerability assessments. It adds two components, viz. 'Non-climatic drivers' and 'Adaptive Capacity'.

The adaptive capacity of society is the capacity to cope with the changes to the external factors like climate. Non-climatic factors determine adaptive capacity of a system or society (Fig. 4.2). Brooks (2003) classifies factors that determine adaptive capacity into hazard specific and generic factors, and into endogenous and exogenous factors. Non-climatic drivers affect relevant non-climatic factors which in turn determine the sensitivity of a system to climate change. In this context, globalization and urbanization are two non-climatic drivers and mitigation also influences non-climatic factors (Fussel and Klein 2006). The relationship between adaptive capacity and adaptation shows that adaptive capacity determines the feasibility of the implementation of adaptation.

4.2   Vulnerability by Security Diagram and Fuzzy Inference System

The Security Diagram is used to measure drought vulnerability in India (Acosta-Michlik et al. 2005). It has three components namely, environmental stress, state susceptibility and crisis probability curves. It depends on both water stress and socio-economic susceptibility. The assumption is made that higher the water stress, the higher the likelihood of crises. At the same time, the higher is the socio-economic susceptibility (i.e. the lower the adaptive capacity), the lower is the stress to cause a crisis. Using the framework of the Security Diagram vulnerability is expressed in the form of $z = f(x, y)$ where z is the function of two explanatory variables say socio-economic susceptibility (x) and water stress (y). The contour line of the Security Diagram is shown in Fig. 4.3. The contour lines away from the origin represents higher vulnerability and vice versa.

The contour lines $z1$, $z2$ and $z3$ are found to be different levels of vulnerability at varying combinations of socio-economic susceptibility x and water stress y. It is also obvious that vulnerability is quite low at z1 and high at z3. The Security Diagrams points out that crisis is likely to occur at say, points between $z2$ and $z3$, where the levels of socio-economic susceptibility are highest or the adaptive capacity to impacts of water stress are lowest. These contour lines are called as "crisis probability curves" (CPC).

The low crisis probability curve (CPCL) and high crisis probability curve (CPCH) correspond to $z2$ and $z3$ respectively in Fig. 4.4. The probability curves are used as yardstick for measuring the degree of vulnerability of the state over time.

Fig. 4.3 Contour lines of the security diagram. *Source*: Acosta-Michlik et al. 2005

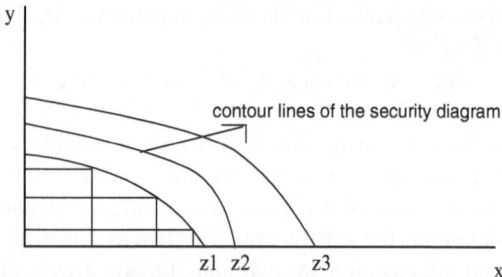

Fig. 4.4 Contour lines of the security diagram. *Source*: Acosta-Michlik et al. 2005

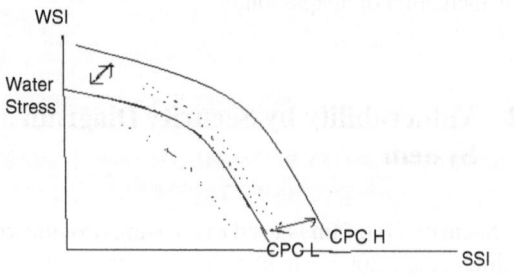

4.2.1 Fuzzy Inference System

Fuzzy Inference System is used to measure drought vulnerability (Bhattacharya and Das 2007; Acosta-Michlik et al. 2005). Fuzzy set theory is useful to translate linguistic statements such as 'high' or 'low' into numerical values. A fuzzy set is the set of real numbers characterized by a membership function in the interval (0, 1). The degree of membership lies between zero and unity. Membership function may be of trapezoidal, triangular, bell-shaped and others.

The 'Low' and 'Very High' vulnerabilities are defined by trapezoid membership functions while the 'Moderate' and 'High' vulnerabilities are defined by triangle membership functions. Intrinsic vulnerability was scaled arbitrary from 1 to 100.

Trapezoidal membership function is given by

$$\text{Trapezoidal}(x : a,b,c,d) = \begin{cases} 0, x < a \\ \dfrac{x-a}{b-a}, a \le x < b \\ 1, b \le x < c \\ \dfrac{d-x}{d-c}, c \le x < d \\ 0, x \ge d \end{cases}$$

Triangular membership function is given by

$$\text{Triangle}(x : a,b,c) = \begin{cases} 0, x < a \\ \dfrac{(x-a)}{(b-a)}, a \le x \le b \\ \dfrac{(c-x)}{(c-b)}, b \le x \le c \\ 0, x > c \end{cases}$$

4.3 Socio-economic Vulnerability for Drought, Coastal and Mountain Areas of West Bengal

To study the socio-economic vulnerability, we have considered six factors like public health facilities, sanitation, educational status; live stock assets, food sufficiency from agriculture and awareness to climate change have been considered from each village. Vulnerability Indices have been calculated using Three Categorized Ranking Method (TCR) assigning scores of 1–3, 1 being the least vulnerable (Shrestha et al. 2003). The basic assumptions are the following;

First, lower level of educational facilities is associated with higher vulnerability
Second, lower level of sanitation is associated with higher vulnerability
Third, higher level of livestock assets is associated with lower vulnerability
Fourth, lower level of awareness to climate change is associated with higher vulnerability
Fifth, higher food insufficiency is associated with higher vulnerability
Sixth, higher health care facility is associated with lower vulnerability.

On the basis of the above assumptions the socio-economic vulnerability are shown in Tables 4.1, 4.2, 4.3 for drought areas, coastal areas and mountain areas respectively. Vulnerability ≥2 means high vulnerability and Vulnerability <2 shows low vulnerability.

It is found from Tables 4.1, 4.2 and 4.3 that the key vulnerabilities are identified in drought, coastal and mountain regions of West Bengal are inaccessibility of education, public health facilities, and inaccessibility of sanitation and food insufficiency. It is also observed from the above tables that the socio-economic vulnerability is higher among the villages in drought and coastal areas of West Bengal compared to socio-economic vulnerability in the mountain areas.

Table 4.1 Socio-economic vulnerability of two villages of drought areas in West Bengal

Village	Education	Sanitation	Livestock assets	Climate awareness	Food sufficiency <3 months	Health care facility	Combined	Vulnerability
Junsura	3	2	3	1	3	3	2.5	H
Baskula	3	2	3	1	3	3	2.5	H

Source: Field Survey
Note: H stands for high, M stands for medium and L stands for low

Table 4.2 Socio-economic vulnerability of two villages of coastal Sunderbans in West Bengal

Village	Education	Sanitation	Livestock assets	Climate awareness	Food sufficiency <3 months	Health care facility	Combined	Vulnerability.
Jamespur	1	3	3	1	3	3	2.3	H
Chargheri	1	3	3	1	3	3	2.3	H

Source: Field Survey
Note: H stands for high, M stands for medium and L stands for low

Table 4.3 Socio-economic vulnerability of two villages of mountain area in West Bengal

Village	Education	Sanitation	Livestock assets	Climate awareness	Food sufficiency <3 months	Health care facility	Combined	Vulnerability
Khoirajhora forest basti	1	1	1	1	1	3	1	L
Rongtong (2)	1	1	1	1	3	3	1.67	L

Source: Field Survey

Note: H stands for high, M stands for medium and L stands for low

Chapter 5
Analysis of Perception, Livelihood Impacts and Threats of Climate Change

This chapter addresses climate change perception index, sea level rise index, livelihood impact and major threats of climate change among the forest dependent people in different agro-climatic regions of West Bengal.

5.1 Climate Change Perception and Climate Change Perception Index of the Households in Drought Prone Area

Perception of climate change in drought areas of West Bengal is analyzed with the help of 11 indicators like realization about longer duration of summer, feeling more warming days in summer than before, feeling the period of winter season has been decreasing, feeling the late starts of winter period, notice any unusual formation of fog, notice the characteristics of one season have been falling in another season. (i.e. overlapping of seasons), observing irregular pattern of rainfall, observing more incidents of drought now a days than before, observing more incidents of storm than before. These 11 indicators have been used to test the perception of climate change.

We have found that more than 90% households have self realization about longer duration of summer in the village of Junsura and Baskula (Table 5.1). More than 96% households have self realization about warmer summer in both the villages. Ninty-five percentage of households have self realization about shorter period of winter in both the villages. Ninty percentage of households have self realization about less cool winter. Most of them have expressed their views that they do not know about "unusual formation of fog, overlapping of seasons" and about 'irregular rainfall pattern' (Table 5.1).

As the forest dependent people have long attachment with the local environment they have experience about the climatic variation. To measure climate experience

© The Author(s) 2017

J.P. Basu, *Climate Change Adaptation and Forest Dependent Communities*,
SpringerBriefs in Environmental Science, DOI 10.1007/978-3-319-52325-5_5

Table 5.1   Climate change perception of the households in Drought prone areas

	Indicators of perception	Self realization	Heard from others	No idea
Junsura	Longer duration of summer	59(98.33)	0.00	1(1.67)
	Warmer summer	58(96.67)	0.00	2(3.33)
	Shorter winter	57(95.00)	1(1.67)	2(3.33)
	Less cool winter	54(90.00)	2(3.33)	4(6.67)
	Late starts of winter	33(55.00)	8(13.33)	19(31.67)
	Unusual formation of fog	1(1.67)	10(16.67)	49(81.67)
	Overlapping of seasons	1(1.67)	11(18.33)	48(80.00)
	Irregular rainfall	6(10.00)	11(18.33)	43(71.67)
	Overall low rainfall	16(26.67)	21(35.00)	23(38.33)
	Increase drought	20(33.33)	23(38.33)	17(28.33)
	Increase stormy events	23(38.33)	21(35.00)	16(26.67)
Baskula	Longer duration of summer	59(98.33)	0.00	1(1.67)
	Warmer summer	58(96.67)	0.00	2(3.33)
	Shorter winter	53(88.33)	1(1.67)	6(10.00)
	Less cool winter	39(65.00)	12(20.00)	9(15.00)
	Late starts of winter	10(16.67)	20(33.33)	30(50.00)
	Unusual formation of fog	3(5.00)	15(25.00)	42(70.00)
	Overlapping of seasons	3(5.00)	12(20.00)	45(75.00)
	Irregular rainfall	13(21.67)	7(11.67)	40(66.67)
	Overall low rainfall	41(68.33)	8(13.33)	11(18.330)
	Increase drought	46(76.67)	7(11.67)	7(11.67)
	Increase stormy events	48(80.00)	6(10.00)	6(10.00)

Note: Figures in the parentheses represent percentage
Source: Field Survey 2011

index, households were asked to rate the statements in 3-point scale. The scale ranges from 1 to 3; whereby 1 means "no idea/cannot remember if heard about", 2 and 3 for "heard from others" and "Observed/felt by own" respectively (Table 5.2). Climate change experience index is calculated on the basis of the indicators stated in Table 5.2 in the drought prone areas of West Bengal. The value of overall reliability coefficient (α) is 0.737.

5.2   Climate Change Perception, Climate Change Perception Index and Perception Regarding Sea Level Rise Index of the Households in Coastal Area

Perception of climate change in coastal areas of West Bengal has been assessed by the same 11 indicators as described earlier. It has been found that the households have been noticing climatic change. Most of the households perceive climate

Table 5.2 Climate Experience Index in drought prone areas

Experience of climate variation/weather anomaly (climate experience index)		
Scale of measurement:	Indicator	Reliability of response
3 = Felt/Observed by own	Respondents' experience with resent frequent:	Alfa (α) if item deleted
2 = Heard from others	Longer (duration of) summer	0.989
1 = Cannot remember if heard about/felt or observed	Warmer summer	0.978
	Shorter (duration of) winter	0.950
	Less cool winter	0.889
	Very late starts of winter	0.650
	Unusual formation of fog	0.425
	Overlapping of seasons	0.419
	Erratic rainfall	0.489
	Overall low rainfall	0.731
	Increase drought event	0.783
	Scarcity of irrigation	0.803
Total responses (N = 120)	Overall Reliability Coefficient α	0.737

Note: Overall reliability coefficient α, is greater than the acceptable limit of 0.70 (Hair et al. 2006)

change. They reported that they have been realizing the increased period of summer, feel warmer in summer etc. (Table 5.3). In addition, climate change experience index and sea-level rise index have been calculated in the two coastal villages. The value of overall reliability coefficients for climate change experience index and sea-level rise index was 0.812 and 0.940 respectively (Tables 5.4 and 5.5). These two values of the said indices are greater than the acceptable limit of 0.70 (Hair et al. 2006).

5.3 Climate Change Perception and Climate Change Perception Index of the Households in Mountain Area

In the mountain regions of West Bengal most of the households have perceived climate change (Table 5.6). The same 11 indicators have been used to access the climate change perception of the households. The value of overall reliability coefficients for climate change experience index is 0.80 (Table 5.7).

Table 5.3 Climate change perception of the households in coastal areas

	Indicators of perception	Self realization	Heard from others	No idea
Jamespur	Longer duration of summer	80(77)	13(13)	11(10)
	Warmer summer	95(91)	8(8)	1(1)
	Shorter winter	71(68)	11(11)	22(21)
	Less cool winter	68(65)	7(7)	29(28)
	Late starts of winter	42(40)	24(25)	38(35)
	Unusual formation of fog	47(45)	18(17)	39(38)
	Overlapping of seasons	38(37)	22(21)	44(42)
	Irregular rainfall	63(61)	24(23)	17(16)
	Overall low rainfall	79(76)	10(10)	15(14)
	Increase stormy events	85(81.7)	10(9.6)	9(8.7)
Chargheri	Longer duration of summer	87(89)	8(8)	3(3)
	Warmer summer	90(92)	5(5)	3(3)
	Shorter winter	84(86)	11(11)	3(3)
	Less cool winter	90(92)	5(5)	3(3)
	Late starts of winter	67(68)	12(12)	19(20)
	Unusual formation of fog	19(20)	5(5)	74(75)
	Overlapping of seasons	19(20)	10(10)	69(70)
	Irregular rainfall	61(62)	19(20)	18(18)
	Overall low rainfall	58(59)	22(22)	18(19)
	Increase stormy events	81(83)	12(12)	5(5)

Note: Figures in the parentheses represent percentage
Source: Field Survey

Table 5.4 Climate experience index in coastal areas

Experience of climate variation/weather anomaly (climate experience index)		
Scale of measurement:	Indicator	Reliability of response
3 = Self realization	Respondents' experience with resent frequent:	Alfa (α) if item deleted
2 = Heard from others	Longer duration of summer	0.919
1 = No idea	Warmer summer	0.965
	Shorter duration of winter	0.881
	Less cool winter	0.875
	Very late starts of winter	0.752
	Unusual formation of fog	0.589
	Overlapping of seasons	0.571
	Erratic rainfall	0.814
	Overall low rainfall	0.838
	Increase stormy events	0.917
Total responses (N = 202)	Overall Reliability Coefficient α	0.812

Note: Overall reliability coefficient α, is greater than the acceptable limit of 0.70 (Hair et al. 2006)

Table 5.5 Perception about sea level rise index in coastal areas

Perception about CC-SLR event (Perception about CC-SLR index)		
Scale of measurement:	Indicator	Reliability of response
3 = Felt/Observed by own	Respondents' perception about:	Alfa (α) if item deleted
2 = Heard from others		
1 = No idea		
	Accelerated sea level rise	0.911
	Rapid/more inward shift of coastline	0.959
	Permanent encroachment of new areas by saline water	0.939
	Acute scarcity of salt free/sweet water for drinking	0.952
Total responses (N = 202)	Overall Reliability Coefficient α	0.940

Note: Overall reliability coefficient α, is greater than the acceptable limit of 0.70 (Hair et al. 2006)

Table 5.6 Climate change perception of the households in the two villages of mountain regions

	Indicators of perception	Self realization	Heard from others	No idea
Khoirajhora forest basti	Longer duration of summer	28(96.5)	0	1(3.5)
	Warmer summer	28(96.5)	1(3.5)	0
	Shorter winter	26(89.6)	1(3.4)	2(7)
	Less cool winter	22(75.8)	0	7(24.2)
	Late starts of winter	13(44.8)	2(7.1)	14(48.1)
	Unusual formation of fog	6(20.6)	3(10.5)	20(68.9)
	Overlapping of seasons	7(24.2)	2(6.9)	20(68.9)
	Irregular rainfall	21(72.4)	1(3.4)	7(24.2)
	Overall low rainfall	24(82.7)	3(10.5)	2(6.8)
	Scarcity of irrigation	26(89.6)	0	3(10.4)
Rongtong (2)	Longer duration of summer	39(92.8)	1(2.3)	2(4.9)
	Warmer summer	39(92.8)	3(7.2)	0
	Shorter winter	36(85.7)	0	6(14.3)
	Less cool winter	33(78.5)	0	9(21.5)
	Late starts of winter	17(40.4)	2(4.9)	23(54.7)
	Unusual formation of fog	9(21.5)	0	33(78.5)
	Overlapping of seasons	6(14.3)	1(2.3)	35(83.4)
	Irregular rainfall	30(71.4)	0	12(28.6)
	Overall low rainfall	39(92.8)	0	3(7.2)
	Scarcity of irrigation	42(100)	0	0

Note: Figures in the parentheses represent percentage
Source: Field Survey 2011

Table 5.7 Climate change experience index in the two villages of mountain regions

Experience of climate variation/weather anomaly (climate experience index)		
Scale of measurement:	Indicator	Reliability of response
3 = Self realization	Respondents' experience with resent frequent:	Alfa (α) if item deleted
2 = Heard from others	Longer (duration of) summer	0.97
1 = No idea	Warmer summer	0.98
	Shorter (duration of) winter	0.92
	Less cool winter	0.85
	Very late starts of winter	0.63
	Unusual formation of fog	0.49
	Overlapping of seasons	0.47
	Erratic rainfall	0.82
	Overall low rainfall	0.94
	Scarcity of irrigation	0.97
Total responses (N = 71)	Overall Reliability Coefficient α	0.80

Note: Overall reliability coefficient α, is greater than the acceptable limit of 0.70 (Hair et al. 2006)

5.4   Livelihood Impact of Climate Change on the Households in Drought Prone Area

This section attempts to examine the impacts of climate change on livelihoods of the forest dependent people. The households were asked to identify the impacts of climate change on their livelihoods in terms of the amount of paddy production either increased or decreased, amount of honey collection either increased or decreased, honey collecting days either increased or decreased, time for the collection of NTFPs either increased or decreased, number of working days in agricultural work either increased or decreased and also the collection amount of fuel wood either increased or decreased. After identifying these impacts, households were ranked according to their responses (yes , no , do not know). Most of them have replied that the production of paddy is decreasing (70% in the village of Junsura and 50% in the village of Baskula) (Table 5.8), 66% in the village of Junsura and 53% in the village of Baskula have replied that the collection of honey has been decreasing, 68% in the village of Junsura and 61% in the village of Baskula have also confirmed about the decreasing number of days for honey collection, time for NTFPs collection has been increasing (50% in the village of Junsura and 58% in the village of Baskula have reported yes). Ninty-three percentage in the village of Junsura and 100% in the village Baskula have reported they do not know about the amount of sal leaves increased or decreased.The number of working days for agricultural labour have been decreasing (50% households in the village of Junsura and 53% in the village of Baskula).

Table 5.8 Impact of climate change on livelihood in drought prone areas of West Bengal

	Perception	Decrease production of paddy	Decrease honey collection	Decrease honey collecting days	Increased time for NTFPs	Decrease amount of Sal leaves	Decrease no. of working days in agriculture	Decrease collection of fuel wood
Junsura	Yes	42(70.00)	40(66.67)	41(68.33)	30(50)	2(3.330)	30(50)	2(3.33)
	No	1(1.67)	4(6.67)	2(3.33)	10(16.7)	2(3.33)	12(20)	5(8.33)
	Don't know	17(28.33)	16(26.67)	17(28.33)	20(33.3)	56(93.33)	18(30)	53(88.33)
	All	60(100)	60(100)	60(100)	60(100)	60(100)	60(100)	60(100)
Baskula	Yes	30(50.00)	32(53.33)	37(61.67)	35(58.4)	0(0.00)	32(53.3)	2(3.33)
	No	10(16.67)	9(15.00)	3(5.00)	5(8.3)	0(0.00)	10	0(0.00)
	Don't know	20(33.33)	19(31.67)	20(33.33)	20(33.3)	60(100.00)	18	58(96.67)
	All	60(100)	60(100)	60(100)	60(100)	60(100)	60(100)	60(100)

Note: Figures in the parentheses represent percentage
Source: Field Survey 2011

Table 5.9 Impact of Climate change on livelihood in coastal areas of West Bengal

	Responses	Fish collection decreasing (%)	Decrease fishing days (%)	Decrease fish catch per go (%)	Honey collection decreasing (%)	Decrease honey collecting days (%)	Crabs collection decreasing (%)
Jamespur	Yes	99(95.2)	90(86.5)	88(84.6)	42(40.4)	41(39.4)	98(94.3)
	No	0(0)	8(7.7)	7(6.7)	20(19.2)	14(13.5)	2(1.9)
	Don't know	5(4.8)	6(5.8)	9(8.7)	42(40.4)	49(47.1)	4(3.8)
	All	104(100)	104(100)	104(100)	104(100)	104(100)	104(100)
Chargheri	Yes	96(98)	93(95)	89(90.8)	9(9.1)	9(9.2)	86(87.8)
	No	1(1)	4(4)	4(4.2)	23(23.6)	21(22.4)	2(2)
	Don't know	1(1)	1(1)	5(5)	66(67.3)	67(68.4)	10(10.2)
	All	98(100)	98(100)	98(100)	98(100)	98(100)	98(100)

Note: Figures in the parentheses represent percentage
Source: Field Survey

5.5   Livelihood Impact of Climate Change on the Households in Coastal Area

Households are very poor and their main source of livelihoods is fishing and crabs collection. They reported that the collection of fish and crabs have been decreasing (Table 5.9) due to the presence of large companies (like Sahara private limited) which collect the fish and crabs with advanced technology in the sea. They also reported that the amount of honey collection has been decreasing (40% in the village of Jamespur and 9% in the village of Chargheri) due to the existence of private enterprises which produce honey artificially with "honey box".

5.6   Livelihood Impact of Climate Change on the Households in Mountain Area

The impact of climate change on livelihoods of the households in mountain regions of West Bengal is presented in Table 5.10. Seventy-two percentage of households in Khoirajhora forest basti confirms the lower is the agricultural production and collection of fuel wood from the forest department has also been decreasing. Thus, in the mountain regions of West Bengal we found a substantial impact of climate change on livelihoods.

5.7   Threats of Climate Change in Drought, Coastal and Mountain Areas of West Bengal

5.7.1   Threats Identified in Drought Prone Areas

Different types of threats realized by the households in the drought prone areas of West Bengal are shown in Table 5.11. The households have ranked their threats on the basis of severity. About 73% households in the village of Junsura and 51% households in the village of Baskula have reported that storm as their first threat

Table 5.10 Impact of climate change on livelihood in mountain areas of West Bengal

		Yes	No	Do not know
Khoirajhora forest basti	Decrease production of paddy	21(72.4)	0	8(27.6)
	decrease amount of fuel wood	26(89.6)	0	3(10.4)
Rongtong (2)	Decrease production of paddy	7(16.6)	0	35(83.4)
	decrease amount of fuel wood	30(71.4)	6(14.3)	6(14.3)

Note: Figures in the parentheses represent percentage
Source: Field Survey 2011

Table 5.11 Relative importance of threats by the households in drought areas

	Threats	First	Second	Third	Fourth
Junsura	Storm	44(73.33)	0(0.00)	0(0.00)	3(5.00)
	Drought	2(3.33)	21(35.00)	2(3.33)	5(8.33)
	Drinking water	3(5.00)	25(41.67)	23(38.33)	0(0.00)
	Health	6(10.00)	10(16.67)	24(40.00)	18(30.00)
	Attack of elephant	5(8.3)	4(6.67)	11(18.33)	34(56.67)
	Crop damage	–	–	–	–
Baskula	Storm	31(51.67)	1(1.67)	2(3.33)	5(8.33)
	Drought	3(5.00)	6(10.00)	4(6.67)	0(0.00)
	Drinking	8(13.33)	24(40.00)	9(15.00)	5(8.33)
	Health	3(5.00)	21(35.00)	26(43.33)	2(3.33)
	Attack of elephant	15(25)	8(13.33)	19(31.67)	46(76.67)
	Crop damage	–	–	–	2(3.33)

Note: Figures in the parentheses represent percentage
Source: Field Survey 2011

while 41% households in the village of Junsura and 40% in the village of Baskula have reported that scarcity of "drinking water" as their second threats. Forty percentage of households in the village of Junsura and 43% households in the village of Baskula have reported "health" hazard as their third threats. Fifty-six percentage of households in the village of Junsura and 76% households in the village of Baskula have expressed that the 'attack of elephant' is their fourth threat. As the villagers of Baskula live in a more deep forest area, they meet more incidents of the "attack of elephant".

5.8   Threats Identified in Coastal Sunderbans

Abnormal climatic conditions due to climatic change have becoming a threat to local people. The households have asked to identify their climatic threats in accordance with severity. It was found that about 44% households in the village of Jamespur and 34% in the village of Chargheri reported that the first threat was flood. About 31% households in the village of Jamespur and 30% in the village of Chargheri reported that cyclone was the second threat (Table 5.12).

Another threat that the villagers often have to face is the attack of Royal Bengal Tiger. Households reported that they have been attacked by Royal Bengal Tiger when they set out for collection of fish and crabs.

Table 5.12   Relative importance of threats by the households in coastal areas

	Threats	First	Second	Third	Fourth
Jamespur	Arrival of tiger	34(32.6)	14(13.4)	32(30.76)	24(23.08)
	Flood	46(44.2)	13(12.5)	6(5.7)	1(0.95)
	Cyclone	18(17.3)	33(31.7)	18(17.3)	10(9.6)
	House damage	1(0.95)	28(26.9)	20(19.2)	25(24)
	Health hazard	5(4.8)	13(12.5)	19(18.2)	36(34.6)
	Damage of agricultural crop	–	3(2.8)	9(8.6)	8(7.7)
Chargheri	Arrival of tiger	53(24.08)	19(19.38)	44(44.9)	12(12.44)
	Flood	04 (34.08)	01(1.02)	02 (2)	05(5.1)
	Cyclone	34 (30.7)	30(30.6)	16(16.32)	25 (25.5)
	House damage	03(7.06)	26(26.53)	17(17)	13(13)
	Health hazard	04(4)	9(9)	19(19)	43(43)
	Damage of agricultural crop	–	–	–	–

Note: Figures in the parentheses represent percentage
Source: Field Survey

Households suffer from various diseases like fever, skin problem, and vision problem and many other heat and water-related diseases. Households cited health hazard as their fourth threat (34% in the village of Jamespur and 43% in the village of Chargheri reported).

5.9   Threats Identified in Mountain Regions

Geographically Darjeeling district is divided into two broad divisions, one is hills and other the plains. The hill area is formed of comparatively recent rock structure that has a direct bearing on landslides. However, heavy monsoon precipitation is a very common cause to the landslides. Two villagers are facing different threats like storm, problem of drinking water, health problem, attack of elephant, damage of crops, landslides etc. The households in the two villages were asked to provide their answer on threats that they face in connection with the climate change. The first threat that the households in the village Khoirajhora forest basti faced is the attack of elephant followed by health hazard. On the other hand the first threat in the village of Rongtong (2) is landslides followed by health hazard and water for drinking purposes. About 71% households reported that their first threat was landslides in village Ronhgton (2) (Table 5.13).

Table 5.13 Relative importance of threats by the households in mountain regions

	Threats	First	Second	Third	Fourth
Khoirajhora forest basti	Storm	0	3(10.34)	5(17.24)	15(51.72)
	Drought	0	0	0	0
	Drinking	0	6(20.68)	4(13.79)	5(17.24)
	Health	0	10(34.48)	14(48.27)	6(20.68)
	attack of elephant	29(100)	0	0	0
	House damage	0	3(10.34)	4(13.79)	1(3.44)
	Damage of crops	0	4(13.79)	2(6.89)	2(6.89)
	Land slides	0	3(10.34)	0	0
Rongtong (2)	Storm	6(14.28)	7(16.7)	8(19.04)	21(50)
	Drought	0	0	0	0
	Drinking	0	3(7.14)	7(16.7)	8(19.04)
	Health	0	8(19.04)	13(30.95)	10(23.8)
	attack of elephant	5(11.90)	20(47.6)	12(28.57)	3(7.14)
	House damage	1(2.4)	0	0	0
	Damage of crops	0	1(2.4)	0	0
	Land slides	30(71.42)	3(7.14)	2(4.8)	0

Note: Figures in the parentheses represent percentage
Source: Field Survey 2011

Chapter 6
Adaptation to Climate Change and Its Determinants at the Household Level

The purpose of this chapter is to identify the adaptation options of the households to manage or to minimize climate risk and to determine the factors responsible for adaptation.

6.1 Adaptation Options of the Households in Drought, Coastal and Mountain Areas of West Bengal

6.1.1 Adaptation to Climate Change in Drought Prone Area

Adaptation is simply ability to cope with the adverse effect of climate change. Households take different adaptive strategies to manage or to minimize climatic risks. Forest dependent people first build up adaptive capacity to minimize risk such as rearing livestock, accessing to NTFPs, forming SHGs, migrating for higher wage and borrowing loan from money lenders etc. These strategies help them to enhance income to build adaptive capacity for climatic risks.

Livestock rearing The households choose mainly cows, goats, hens and pigs that help to maximize net revenue subject to the prices, climate, soils and other external factors that households face. We have found that the households rear five major types of livestock like cows, goats, hens and pigs in the two villages. The major livestock products sold are milk, cow dung, goats, hens, eggs and pigs.

Self Help Groups (SHGs)

Formation of Self help groups (SHGs) has a vital contribution to the forest dependent people to manage their livelihoods in adverse climatic situation. It offers loan which the poor villagers use to build up their assets, increase their wealth, and

© The Author(s) 2017
J.P. Basu, *Climate Change Adaptation and Forest Dependent Communities*,
SpringerBriefs in Environmental Science, DOI 10.1007/978-3-319-52325-5_6

Table 6.1 Different
adaptation strategies in the
two villages under drought
areas

Adaptation strategies	Junsura	Baskula
Livestock rearing	60(100.00)	59(98.33)
Formation of SHGs	23(38.33)	7(11.67)
Accessibility to NTFPs	60(100.00)	59(98.33)
Migration	34(56.67)	46(76.67)
Borrow loan from money lenders	39(65.00)	57(95.00)

Note: Figures in the parentheses represent percentage of total
sample households
Source: Field Survey 2011

enable for starting small business enable to fight against risks and poverty. Key
feature of SHGs programs is that they are mostly directed towards women. Moreover,
it has been found that women, being more credit constrained than men, are more
likely to engage in making sal dish, collecting fuel wood regularly. We have found
that 38% households in the village of Junsura and only 11% households in the village of Baskula have formed SHGs. (Table 6.1).

Non-Timber Forest Products (NTFPs)

The collection of NTFPs is widespread in the two villages. NTFPs have a dominant
role to overcome climatic risk for the forest dependent people. NTFPs collected are
classified as Sal leaves (for making sal dish), *Kutchi* kathi (used for stitching sal
leaves), mushroom, different roots of the plants (used for home medicine and sometimes for selling), fuel wood (for households' consumption and selling), *Kendu pata*
(used for *bidi* making) etc. These are the main forest products collected by the
household from the forest. Households sell these forest products to the local market
for income. Our study has shown that 100% households in the village of Junsura and
98% households in the village of Baskula are engaged to access NTFPs (Table 6.1).

Migration: In our study of the two village 56% households in the village of
Junsura and 76% households in the village of Baskula migrate (temporary and
return to the village) in search of jobs to the agricultural and industrial developed
districts like Burdwan, Nadia and Hooghly.

Borrowing loan from Money lender: Households borrow money from the
money lenders when they need money in some emergency purposes like medical
and drought etc. 65% households in the village of Junsura and 95% households in
the village of Baskula have borrowed loan from money lenders (Table 6.1, Fig. 6.1).

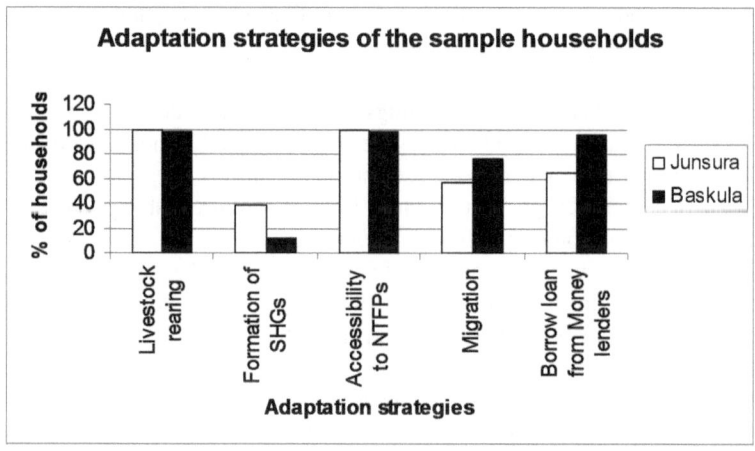

Fig. 6.1 Adaptation strategies of the sample households in drought areas in West Bengal. *Source*: Field Survey 2011

Table 6.2 Different adaptation strategies in the two villages under Coastal Sunderbans

Strategy	Jamespur (Yes)	Chargheri (Yes)
Accessibility of loan from		
1. Banks	12(11.5)	16(16.3)
2. Money lenders	91(87.5)	81(82)
Livestock rearing	82(78.8)	69(70)
SHG	81(77.9)	89(90)
Migration	76(73)	84(85)
Fishing and crabs collection	89(85.5)	89(90)
Diversification into wage labour from agriculture and forestry	80(76.9)	96(98)
NREGA[a]	11 days per year	11 days per year

[a]*Note*: Average no of working days in NREGA
Figures in the parentheses represent percentage
Source: Field survey

6.1.2   Adaptation in Coastal Region

The adaptation strategies in the two villages of coastal Sunderbans are presented in Table 6.2. **Accessibility of loan**: Households access loan for the needs of emergency purpose mainly from money lenders (Table 6.2). They use loan amount generally in unproductive purposes like house repairing, marriage of their daughter, etc. A few use the loan amount in productive purpose like starting small business or buying fishing equipments. (87% in the village of Jamespur and 82% in the village of Chargheri reported that they borrowed loan from money lenders).

   Livestock rearing: Livestock rearing is a tool of income generation of the poor. 78% households in the village of Jamespur and 70% households in the village of

Chargheri reported that they rear live stock asset like cow, goat, hen, sheep and pigs for additional income generation (Table 6.2).

SHGs: Based on the concept of "self-help," small groups of women have formed into groups of 10–20 and operate a savings-first business model whereby the member's savings are used to fund loans. The loans that SHG members receive are intended to improve their livelihoods so that they can receive greater and steadier cash flows. In rural areas, livelihoods range from agriculture, fishing, livestock rearing, dairy and various other goods and services activities.

SHGs members improve their livelihood as well as the strength of income. Increased income helps them to manage climatic risks (77% households in the village of Jamespur and 90% households in the village of Chargheri formed SHG (Table 6.2).

Migration: In the region, migration generally occurs during the non-harvest season when villagers look for additional work elsewhere (73% in the village of Jamespur and 85% in the village of Chargheri reported that they migrate) (Table 6.2). In this region people migrate to Kolkata, urban town, for getting jobs and this is temporary migration.

Fishing and crab collections: The major sources of livelihood of Sunadarbans are fishing and crab collections. The number of fishermen increased due to population growth and emergence of Bagda (tiger shrimp) farming. During Bagda post-larvae collection season nearly 30–40 thousand boat goes to sundarbans for shrimp fry collection. In our study of Jamespur and Chargheri villages (of Gossaba block), it is found that most fishermen are poor and they harvest for subsistence (85% in the village of Jamespur and 90% in the village of Chargheri) (Table 6.2).

Diversification: In the region there is a diversification of works from fishing to wage labour. The main causes behind this diversification include a decrease in both the fish stocks, which are aggravated by climate change like precipitation and temperature rise and increased cyclone. In the region one of the most common adaptations is a diversification of daily wage (76% in the village of Jamespur and 98% in the village of Chargheri reported that they adapting to daily wage labour work for their livelihood) (Table 6.2, Fig. 6.2). The causes of diversification in this region are lower amount of fish and crabs collection and the problems of getting boat licenses etc.

6.1.3   Adaptation in Mountain Region

The adaptation strategies in the two villages of Mountain regions are shown in Table 6.3. Households reported that they have been diversifying their occupation from agriculture to daily wage labour (Table 6.3). As there is no agricultural cooperative or any other banking facility in the region households borrow money from money lenders when landslides, heavy rainfall and droughts occur. About 51% households in the village of Khoirajhora forest basti and 45% in the village of Rongtong (2) reported that they borrowed money from money lenders (Table 6.3).

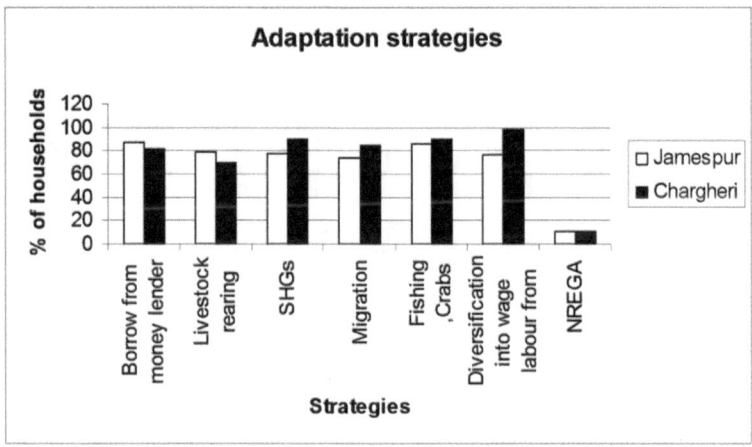

Fig. 6.2 Adaptation strategies of the sample households in Coastal Sunderbans in West Bengal

Table 6.3   Different adaptation strategies in the two villages under Mountain areas

Adaptation strategies	Khoirajhora forest basti	Rontong (2)
Diversification from agriculture to daily wage labour	29(100)	38(90.47)
Borrow money from money lender	15(51.72)	19(45.23)
Member of SHG	–	20(47.6)
Migration	20(68.96)	27(64.28)
Treatment of water	5(70.24)	14(33.33)
Animal husbandry	21(72.41)	27(64.28)

Note: Figures in the parentheses represent percentage of total sample households
Source: Field Survey 2011

Initiatives for SGHs formation has not yet been taken properly in the two studied villages. 47.6% households have formed SHGs in the village Rongtong (2).

Villagers generally get job in and around their locality for 6 months. They migrate to Darjeeling town and Nepal for jobs. About 68.96% households in the village of Khoirajhora forest basti and 64.28% households in the village of Rongtong (2) migrate every year (Table 6.3).

Health problem is a main issue in the study area. People suffer a lot as there is no public health care in that area. People have to travel 7–8 km for accessing health services. They maintain some family precaution to avoid disease. About 70.24% households in the village of Khoirajhora forest basti and 33.3% households in the village of Rongtong (2) reported that they use boiled water for children and patients at home (Table 6.3).

Animal husbandry is a main source of income generation to the people in the study areas. Animal husbandry is a way to minimize agricultural loss and for extra income generation. Villagers rear cows, goats, hens, and earn extra income from home. About 72.41% households in the village of Khoirajhora forest basti and

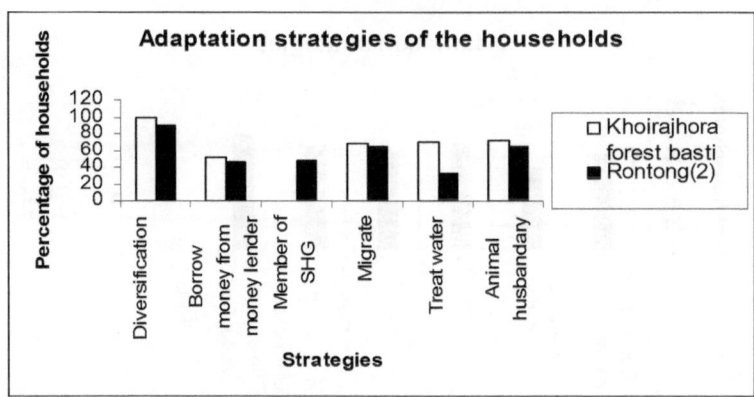

Fig. 6.3 Adaptation strategies of the sample households in mountain region in West Bengal

64.28% households in the village of Rongtong (2) reported that they earn extra income from animal husbandry (Table 6.3, Fig. 6.3).

6.2   Determinants of Adaptation at the Household Level

6.2.1   Determinants of Adaptation in Drought Prone Areas

First we consider the determinants of adaptation in drought prone areas of West Bengal. We have used Heckman Two Stage model to identify the determinants of adaptation to climate change.

Adaptation to climate change is a two-stage process involving perception and adaptation stages. The first stage is whether the respondent perceived there was climate change or not, and the second stage is whether the respondent adapted to climate change on the condition that first stage that he/she had perceived climate change. We have used maximum likelihood Heckman's two-step procedure (Heckman 1976) to correct for the selectivity bias.

6.2.1.1   Description of the Variables in the Model

The variables hypothesized as affecting perceptions and adaptations to changes in climatic conditions along with their respective dependent variables are indicated below.

6.2.1.2 Dependent Variables for the Outcome Equation

In terms of annual income generation we have chosen the income from non-timber forest products as the dependent variable for the outcome model.

6.2.1.3 Explanatory Variables for the Outcome Equation

As indicated in Table 6.4 below, the explanatory variables for this study include: age of the head of the households, marital status, operational holdings, physical asset value, livestock asset value, farm income, wage income, forestry income, temperature and family size.

6.2.1.4 Dependent Variable for the Selection Equation

The analyses of the perception of the forest dependent communities to climate change indicate that most of them for this study are aware of the fact that temperature is increasing.

6.2.1.5 Explanatory Variables for the Selection Equation

For the selection equation, it is hypothesized that, education, age of head of the household, marital status, adult male in the family, operational holdings, physical asset value, livestock asset value and family size influence the awareness of the people to climate change.

The factors affecting the perception and adaptation models are shown in Table 6.4. The Heckman Probit model was run and tested for its appropriateness over the standard probit model. The results indicated that the likelihood function of the Heckman probit model was significant (Wald $\chi^2 = 80.75$, with $p < 0.0000$) showing strong explanatory power of the model. The results from regression indicated that most of the explanatory variables affected the probability of adaptation as expected. Variables that positively and significantly influenced the adaptation to climate change include the age of the household, farm income, forestry income, temperature and family size (Table 6.5).

Age is positively influences the decision to undertake adaptation because elder people are more experienced and have better access to non-timber forest products than younger ones, and hence the former have a higher probability of adopting the practice.

Family size also influences the decision to adapt. There is a possibility that the households with many family members may be forced to collect forest products to ease the consumption pressure imposed by a large family size.

Table 6.4 Description of model variables for the Heckman Probit model

Outcome equation (Adaptation model)			Selection equation (Perception model)		
Dependent variable			Dependent variable		
Description	People reported to have adapted (%)	People reported not adapted (%)	Description	People perceived change in temperature (%)	People not perceived change in temperature (%)
Accessibility of non-timber forest products	96	04	Perception of temperature increased	93	07
Independent variables			Independent variables		
Description	Mean	Standard deviation	Description	Mean	Standard deviation
Age (in years)	42.35	13.06	Age (in years)	42.35	13.06
Education (in years)	1.33	2.1793	Education (in years)	1.33	2.1793
Marital status (Yes = 1, No = 0)	0.98	0.12855	Marital status (Yes = 1, No = 0)	0.98	0.12855
Operational holdings (in acres)	0.0268	0.09607	Adult male (in number)	1.3	0.74020
Physical asset value (in rupees)	2407	1486.291	Operational holdings (in acres)	0.0268	0.09607
Livestock asset value (in rupees)	7454	6452.24	Physical asset value (in rupees)	2407	1486.291
Wage income (in rupees)	6216	2245.242	Livestock asset value (in rupees)	7454	6452.24
Forestry income (in rupees)	17,055	5910.35			
Family size (in number)	3.44	1.1867			
Temperature (in degree centigrade)	41.36	1.4872			

Source: Field survey

Adaptation to climate change increases with increasing temperature. The increasing temperature has damaging effect on agriculture and raises food insecurity. They respond to this through the adoption of different adaptation methods.

Income from forestry has significant and positive impact on adaptation. With higher income from forestry there is a possibility to enhance adaptation in order to minimize the risk of climate change. There is a negative association between operation holdings and adaptation. This means that the low holding farms have greater

Table 6.5 Results of the Heckman Probit selection model

	Estimated coefficients outcome equation: adaptation model (Accessibility of non-timber forest products)		Estimated coefficients selection equation: perception model (Perception of temperature increased)	
	Regression		Regression	
Explanatory variables	Coefficients	P-level	Coefficients	P-level
Age	0.006519[a]	0.041	0.0063211[b]	0.067
Education	−0.064523	0.457	−0.135648	0.458
Marital status	−0.512364[b]	0.069	2.3684	0.647
Adult male			1.64782[a]	0.026
Operational holdings	−0.17568[b]	0.068	1.68241[c]	0.001
Physical asset value	−0.0000179[a]	0.019	0.000254	0.648
Livestock asset value	−4.56e^{-05}	0.648	−0.0000897	0.237
Wage income	−0.0000478[b]	0.079		
Forestry income	0.0000387[c]	0.001		
Temperature	0.0785442[a]	0.018		
Family size	0.0679428[a]	0.041		
Cons	−3.23567[b]	0.054	−8.47512	0.254
Total observations	120			
Censored observations	70			
Uncensored observations	50			
Wald chi square (zero slopes)	80.75[c]	0.000		

[a]Significant at 5% level
[b]Significant at 10% level
[c]Significant at 1% level
Source: Field survey

adaptation compared to the large holding farms. The negative association is also true in the case of physical asset value and wage income. These findings are contrary to the case of the adaptation of agricultural farmers. Variables, say age, the numbers of adult male and operational holdings are found to have significant and positive impact on the perception of temperature increase (Table 6.5).

6.2.2   Determinants of Adaptation in Coastal Areas of Sunderbans

6.2.2.1   Dependent Variables for the Outcome Equation

In the coastal areas of Sunderbans the households' adaptation options are identified as migration, formation of Self-help Group (SHGs), catching fish and crabs, borrowing credit from moneylenders and livestock rearing. In terms of annual income generation we have chosen the fishing and crab collection as dependent variable for the outcome model.

6.2.2.2   Explanatory Variables for the Outcome Equation

As indicated in Table 6.6 below, the explanatory variables for this study include: age of the head of the households, education, marital status, operational holdings, sex, family size, physical asset value, livestock asset value, wage income, income from fishing and crabs, and permanent existence of saline water in the area.

6.2.2.3   Dependent Variable for the Selection Equation

The analyses of the perception of the coastal communities to climate change indicate that most of them for this study are aware about the sea level rise. The perception of sea level rise is the dependent variable for the selection equation.

6.2.2.4   Explanatory Variables for the Selection Equation

For the selection equation, it is hypothesized that, education, age of head of the household, marital status, adult male in the family, operational holdings, education, physical asset value, livestock asset value and family size influence the awareness about sea level rise to climate change.

From Table 6.7 it is observed that there is a positive and significant relation between adaptation to climate change and physical asset value, income from fishing and crabs and permanent existence of saline water. The negative relation is observed in the case of wage income and operational holdings. Variables like education and sex also affect the perception of sea level rise in coastal areas.

Table 6.6 Description of model variables for the Heckman Probit model in coastal Sunderbans

Outcome equation (Adaptation model)			Selection equation (Perception model)		
Dependent variable			Dependent variable		
Description	People reported to have adapted (%)	People reported not adapted (%)	Description	People perceived rise in sea level (%)	People not perceived rise in sea level (%)
Fishing and crab collection	89	11	Perception of Sea Level Rise	96	4
Independent variables			Independent variables		
Description	Mean	Standard deviation	Description	Mean	Standard deviation
Age (in years)	44	13.36	Age (in years)	44	13.36
Education (in years)	4.08	3.40	Education (in years)	4.08	3.40
Maritial status (Yes = 1, No = 0)	0.99	0.070	Maritial status (Yes = 1, No = 0)	0.99	0.070
Operational holdings (in acres)	0.186	0.365	Adult male (in number)	1.44	0.653
Physical asset value (in rupees)	5403	6352	Operational holdings (in acres)	0.186	0.365
Livestock asset value (in rupees)	2402	3547	Physical asset value (in rupees)	5403	6352
Wage income (in rupees)	6841	3901			
Income from fishing and Crab (in rupees)	8927	6010			
Family size (in number)	4.33	1.42			
Sex (Male = 1, Female =0)	0.960	0.155			
Permanent existence of saline water in the area (Yes =1, No = 0)	0.933	0.245			

Source: Field survey

Table 6.7 Results of the Heckman Probit selection model in coastal Sunderbans

Explanatory variables	Estimated coefficients outcome equation: adaptation model (Accessibility of fishing and crab collection		Estimated coefficients selection equation: perception model (Perception of sea level rise)	
	Regression		Regression	
	Coefficients	P-level	Coefficients	P-level
Age	0.0003472	0.807	0.0151184	0.200
Sex	−0.0490691	0.798	−5.25678[a]	0.000
Education	0.0040679	0.25	0.1004056[b]	0.029
Maritial status	−0.6586135	0.100	0.00567	0.87
Adult male			0.2386922	0.327
Operational holdings	−0.1556393[b]	0.021	−0.2173399	0.556
Physical asset value	0.000000743[b]	0.023	−000000268	0.902
Livestock asset value	−000000196	0.771	.0000158	0.732
Income from fishing and crabs	0.0000234[a]	0.000		
Wage income	−0.0000175[a]	0.001		
Family size	0.0052145	0.510		
Permanent existence of saline water in the area	0.527202[c]	0.073		
Cons	0.6273712[a]	0.001		
Total observations	202			
Censored observations	18			
Uncensored observations	184			
Wald chi square (zero slopes)	1364.24	0.0000		

[a]Significant at 1% level
[b]Significant at 5% level
[c]Significant at 10% level
Source: Field survey

Chapter 7
Government's Policy, Conclusions and Recommendations

For developing countries like India, adaptation requires assisting the vulnerable population during adverse climate conditions and empowering them to cope with climate risks in the long-run for better living. This chapter summarizes the findings and a conclusion discussed in Chapters 3–6, discusses government policies to reduce vulnerability and offers recommendations.

7.1 Government's Policy

In the late 1970s local communities began protesting against indiscriminate destruction of the forests and they launched a movement called Chipko movement of the Himalayas (Hedge 1998). India's National Forest Policy of 1988 has also recognized the interdependence between people and forests, and envisages active community participation in the protection and development of forestlands for sustainability of forest management (Sarker and Das 2008). The government of India has adopted community forestry in the name of Joint Forest Management (JFM) as a principal approach for community-based forestry in 1990. Under Joint Forest management (JFM) the effective involvement of village communities in evolving sustainable forest management systems has been considered as an important approach to address the long-standing problems of deforestation and land degradation in India. The involvement and participation of local communities in climate change reduction programs took the highest priority, since they are the first victims to climate change event. In India, Joint Forest Management committees (JFMCs) have been set up at the village level to involve people in forest protection and conservation. There are 1, 18,213 JFM committees covering 29 States in India. JFMCs are also responsible for protecting the forest from fires (Singh et al. 2007). This community based management institutions often considered as a critical precondition for equitable, efficient and effective implementation of REDD+ (Springate-Baginski and Wollenberg 2010). The Joint Forest Management committees (JFMCs)

© The Author(s) 2017   71
J.P. Basu, *Climate Change Adaptation and Forest Dependent Communities*,
SpringerBriefs in Environmental Science, DOI 10.1007/978-3-319-52325-5_7

are also involved in reducing illicit felling of trees, reducing seized timber and protecting forest from illegal loggings in West Bengal (West Bengal State forest report 2006–2007).

Mission for a Green India known as the National Mission for Green India under the National Action Plan on Climate Change (NAPCC) acknowledges that the forestry sector has multi functional role in influencing climate mitigation, food security, water security, biodiversity conservation and livelihood security of forest dependent communities. There are three objectives of the Mission. First is to increase forest cover of 10 million ha out of which 5 million ha of forest/non-forest land used for forest/tree cover and rest 5 million ha for the quality of forest cover. Second is the improvement of ecosystem services like biodiversity, hydrological services and carbon sequestration. Third is the generation of forest-based livelihood income of three million households who are living in and around the forests. Fourth is the enhancement of annual CO_2 sequestration from 50 million tonnes to 60 million tonnes in the year 2020.

Local institutions have a significant role for forest conservation and its sustainable use. The institutions at the local level are Joint Forest Management Committees (JFMC), Community Forest Management groups (functioning in Orissa), Van Panchayats (functioning in Uttarakhand), traditional village level institutions or Village Councils (schedule VI area) and Biodiversity Management Committees etc. Self Help Groups also promotes forest-based livelihood activities in the study areas.

The spread of Joint Forest Management, despite several limitations and uncertainties in terms of tenurial insecurity, inadequate silvicultural development, and restricted harvesting and market access, has helped in regenerating forests and 16 meeting local needs (Milne 2006).

Panchayati Raj Institutions (PRIs) are constitutionally mandated bodies for decentralized development planning and execution at the local level (MoEF 2009). The Scheduled Tribes and Other Forest Dwellers (Recognition of Forest Rights) Act, 2006, in addition to individual rights, provides for Community Forest Rights, including the right to protect, regenerate and manage Community Forest Resource.

Social Protection measures are needed to cope with the adverse effects of climate change. In this context Microfinance services (MFS) and Mahatma Gandhi National Rural Employment Guarantee Act in 2005 (MNREGA) are vital for reducing climate change vulnerability. MGNREGA contributes to the reduction of greenhouse gas emissions through plantation afforestation as well as horticulture, land development, well construction, renovation of ponds and production of biomass and wood and also carbon sequestration.

Micro finance services (MFS) plays a key role to act as climate change adaptation. MFS has potential to help the most vulnerable populations by providing financial services to become less susceptible to shocks and stresses and to cope with the impacts of climate change. Microfinance policy links the poor through Micro credit, Micro insurance, Micro savings policy which help them to enhance income.

The Government of India implements a series of central and centrally sponsored schemes under different ministries and departments for achieving social and economic development. At present, all of the schemes are not related to adaptation schemes but many contain elements (objectives and targets) that clearly relate to risks from climate variability. A recent initiative by the Department of International Development (DFID) and the World Bank in India seeks to identify how to integrate adaptation and risk reduction into their portfolio of programs. The programs include National Rural water and Sanitation Program, National Elementary Education Program (Sarva Shiksa Abhiyan), National Reproductive and Child Health Program Phase II, Kolkata Urban Services for the Poor, West Bengal Support to Rural Decentralization, West Bengal Health Systems Development Initiatives, Andra Pradesh Rural Livelihoods Program, Madhya Pradesh Rural Livelihood Program, and Madhya Pradesh Urban Services for the Poor, and Western Orissa Rural Livelihood Project. Besides, the housing scheme, Indira Awas Yojana, the Food for Work Programme, and the rural road building scheme, Pradhan Mantri Grameen Sadak Yojana. These schemes have provided relief in the aftermath of floods and cyclones, enabled recovery and rebuilding, and helped improve connectivity selling produce and finding alternative employment.

7.2 Conclusions

The following conclusions are emerged from the above the study;

First, the socio-economic conditions of the households in drought, coastal and mountain regions of West Bengal are weak. There have been prevailing acute poverty, most of the households are landless, small and marginal farmers and they are illiterate. Most of the sample households are constituted by scheduled castes and scheduled tribes.

Second, most of the households are not taking the advantage of infrastructural facilities like electricity facility, facilities for drinking water, sanitation, public health care and borrowing facilities from banks.

Third, in the drought area the main source of income is the non-timber forest products, in the coastal area most of the households derive their income from fishing and crab collections while in the mountain regions animal husbandry is the main source of income.

Fourth, in the drought area more than 90% households have self realized longer duration of summer in the village of Junsura and Baskula. More than 96% households have self realized warmer summer in both the villages. Ninety-five percent households have self realized shorter period of winter in both the villages. Similarly, 90% households have self realized about less cool winter. Maximum numbers of households have no idea about unusual formation of fog and overlapping of seasons as it is 81% and 80% respectively in both the villages. The value of overall reliability coefficient was 0.737.

Perception of climate change in coastal areas of West Bengal has been assessed by the same 11 indicators described above. It has been found that the households have been noticing climate changes. They reported that they have been realizing the increased period of summer, feel warmer in summer etc. In addition, climate change experience index and sea-level rise index have been calculated in the two coastal villages. The value of overall reliability coefficients for climate change experience index and sea-level rise index was 0.812 and 0.940 respectively. In the mountain regions of West Bengal most of the households perceived the climate change. The value of overall reliability coefficients for climate change experience index was 0.80.

Fifth, there has been an impact of climate change on the livelihood of the sample households. Most of the households in drought areas reported that the production of paddy decreasing (70% in the village of Junsura and 50% in the village of Baskula), 66% households in the village of Junsura and 53% in the village of Baskula have reported that the collection of honey has been decreasing, 68% households in the village of Junsura and 61% in the village of Baskula have also confirmed about the decreasing number of days for honey collection, time for NTFPs collection has been increasing (50% in the village of Junsura and 58% in the village of Baskula have reported yes). The number of working days for agricultural labour have been decreasing (50% households in the village of Junsura and 53% in the village of Baskula have said yes).

We found the impact of climate change on livelihood in the coastal people in West Bengal. Households' main sources of livelihood are fishing and crabs collection, honey collection. They reported that the collection of fish and crabs have been decreasing (99% in the village of Jamespur and 98% in the village of Chargheri). They also pointed out that the amount of honey collection has been decreasing.

In the mountain areas of West Bengal we found a substantial change of livelihood due to climate change. Seventy percent households in Khoirajhora forest basti confirms the lower is the agricultural production (food crop) while the collection of fuel wood from the forest has been decreasing.

Sixth, the key vulnerabilities are identified as lack of education, ill health hygiene and food insufficiency.

Seventh, In the drought prone areas of West Bengal the households' adaptation strategies are identified as collection of non-timber forest products, migration, formation of self-help groups, and livestock is rearing. In the coastal areas of Sunderbans the households' adaptation options are identified as migration, formation of Self-help Group (SHGs), catching fish and crabs, borrowing credit from moneylenders and livestock rearing. In the mountain regions the households' adaptations options are migration, borrowing loan from money lenders, diversifications of occupations from agriculture to wage labour, use boiled water for children at home and animal husbandry.

Eight, the study has identified different types of threats of climate change in different regions of West Bengal. In the drought areas the threats are likely to be storms, shortage of drinking water, heath hazards and attack of elephants. Similarly, in the coastal area the threats are floods, cyclones, attack of tiger, health hazards etc. In the mountain regions the land slides, storm, problem of drinking water, health problem, attack of elephant and damage of crops etc. are the possible threats in connection with the climate change.

Nine, the study examines the determinants of adaptation in drought and coastal areas of West Bengal. In the drought areas variables that positively and significantly influenced the adaptation to climate change include the age of the household, farm income, forestry income, temperature and family size. There is a negative association between operation holdings and adaptation. The negative association is also true in the case of physical asset value and wage income. Variables say age, the numbers of adult male and operational holdings are found to be significant and positive impact on the perception of temperature increased.

In the coastal areas there is a positive and significant relation between adaptation to climate change and physical asset value, income from fishing & crabs and permanent existence of saline water. The negative relation is observed in the case of wage income and operational holdings. Variables like education and sex also affect the perception of sea level rise in coastal areas.

7.3 Recommendations

The study emphasizes the enhancement of adaptive capacity which depends on economic, social, and human development, which in turn related to income, inequality, poverty, literacy, and regional disparity; capacity and governance of public institutions education, health, social protection, and social safety nets. In this context, the study has identified some immediate priorities. First is an integrated water management program like flood control and prevention, early warning flood systems, irrigation improvement, and demand-side management. Second is strengthening local adaptive capacity like better climate information, research and development on heat-resistant crop variety and other techniques, early warning systems, and efficient irrigation systems, and index-based insurance schemes. Third is building up awareness-raising programs, reforestation and afforestation. Fourth is the management of mangrove conservation and planting. Fifth is the introduction of "climate proofing" transport-related investments and infrastructure.

7.3. Recommendations

References

Abeygunawardena, P., and A. Wikramasinghe. 1992. An Economic evaluation of non-timber products of Hantana Forest. In *Paper Presented at the Workshop on Methods for Social Science Research on Non-Timber Forest Products*, May 18–20, 1992, Bangkok, Thailand.

Abson, D.J., A.J. Doughill, and L.C. Stringer. 2012. Using principal component analysis for information rich socio-ecological vulnerability mapping in Southern Africa. *Applied Geography* 35: 515–524.

Acosta-Michlik, Lilibeth, Fausto Galli, Richard J.T. Klein, Sabine Campe, Kavi Kumar, Frank Eierdanz, Joseph Alcamo, Dörthe Krömker, Alexander Carius, and Dennis Tänzler. 2005. How vulnerable is India to climatic stress? In *Measuring vulnerability to Drought Using the Security Diagram Concept. An International Workshop*, Holmen Fjord Hotel, Asker, Near Oslo, 21–23 June.

Adams, R.M. 1989. Global climate change and agriculture: An economic perspective. *American Journal of Agricultural Economics* 71 (5): 1272–1279.

Adams, W.M., R. Aveling, D. Brockington, B. Dickson, J. Elliott, J. Hutton, D. Roe, B. Vira, and W. Wolmer. 2004. Biodiversity conservation and the eradication of poverty. *Science* 306: 1146–1149.

ADB. 2009. *Asian development outlook 2009*. Manila: Asian Development Bank.

Adger, W.N., N. Brooks, G. Bentham, M. Agnew, and S. Eriksen. 2004. *New indicators of vulnerability and adaptive capacity*. Technical Report 7. Norwich: Tyndall Centre for Climate Change Research, University of East Anglia.

Adger, W.N., and K. Vincent. 2005. Uncertainty in adaptive capacity. *Comptes Rendus Geosciences* 337 (4): 399–410.

Adger, W.N. 1999. Social Vulnerability to Climate Change and Extremes in Coastal Vietnam‖. *World Development* 27: 249–269.

———. 2000. Institutional adaptation to environmental risk underthe transition in Vietnam. *Annals of the Association of American Geographers* 90: 738–758.

———. 2006. Vulnerability. *Global Environmental Change* 16: 268–281.

Adger, W.N., N.W. Arnell, and E.L. Tompkins. 2005. Successful adaptation to climate change across scales. *Global Environmental Change* 15: 77–86.

Adger, W.N., and M. Kelly. 1999. Social vulnerability to climate change and the architecture of entitlements. *Mitigation and Adaptation Strategies for Global Change* 4: 253–266.

Amdu, B., Azemeraw Ayehu, and Andent Deressa. 2013. *Farmers' perception and adaptive capacity to climate change and variability in the Upper Catchment of Blue Nile, Ethiopia*. African Technology Policy Studies Network. Working Paper Series No. 77.

© The Author(s) 2017   77
J.P. Basu, *Climate Change Adaptation and Forest Dependent Communities*,
SpringerBriefs in Environmental Science, DOI 10.1007/978-3-319-52325-5

Arnold, J.E.M. 1995. Socio-economic benefits and issues in non-wood forest product use. In *Report of the International Expert Consultation of Non-Wood Forest Products*, 89–123. Rome: Food and Agriculture Organization of the United Nations.

Basu, Gargi. 2015. Vulnerability and adaptation to climate change: A study on forest dependent communities of Indian Sunderban. Ph.D. thesis, West Bengal State University, Barasat, Kolkata, India.

Benayas, J.M.R., A.C. Newton, A. Diaz, and J.M. Bullock. 2009. Enhancement of biodiversity and ecosystem services by ecological restoration: A meta-analysis. *Science* 325: 1121–1124.

Bennett K. 2001. *Voicing power: Women, family farming and patriarchal webs*. Centre for Rural Economy Working Paper Series Working Paper 62, February 2001.

Berkes, F., and C. Seixas. 2006. Building resilience in lagoon social-ecological systems: A local-level perspective. *Ecosystems* 8: 967–974.

Bhattacharya, S., and A. Das. 2007. *Vulnerability to drought, cyclones and floods in India*. BASIC, Paper 9, September 2007.

Bohle, H.G., T.E. Downing, and M.J. Watts. 1994. Climate change and social vulnerability: Toward sociology and geography of food insecurity. *Global Environmental Change* 4 (1): 37–48.

Brooks, N. 2003. *Vulnerability, risk and adaptation: A conceptual framework*. Working Paper 38. Norwich: Tyndall Centre for Climate Change Research, University of East Anglia. http://www.tyndall.ac.uk/publications/working_papers/wp38.pdf.

Brooks, N., and W.N. Adger. 2003. *Country level risk measures of climate-related natural disasters and implications for adaptation to climate change*. Tyndall Centre Working Paper No. 26.

Brooks, N., W. Neil Adger, and P. Mick Kelly. 2005. The determinants of vulnerability and adaptive capacity at the national level and the implications for adaptation. *Global Environmental Change* 15 (2): 151–163. doi:10.1016/j.gloenvcha.2004.12.006.

Byron, R.N., and J.E.M. Arnold. 1999. What futures for the people of the tropical forests? *World Development* 27 (5): 789–805.

Canadell, J.G., and M.R. Raupach. 2008. Managing forests for climate change mitigation. *Science* 320: 1456–1457.

Cardona, O.D. 2005. Indicators of disaster risk and risk management—main technical report. IDB/IDEA Program of Indicators for Disaster Risk Management, National University of Colombia, Manizales. http://idea.unalmzl.edu.co.

Chakraborty, Anusheema, and P.K. Joshi. 2014. Mapping disaster vulnerability in India using analytical hierarchy process. *Geomatics, Natural Hazards and Risk* 7(1): 308–325.

Chandrasekharan, C. 1998. Role of non-wood forest products in sustainable forest management, SEANN Workshop, Dehra Dun.

Chaturvedi, R.K., R. Gopalakrishnan, M. Jayaram, G. Bala, N.V. Joshi, R. Sukumar, and N.H. Ravindranath. 2010. Impact of climate on Indian forests: A dynamic vegetation modeling approach. *Mitigation and Adaptation Strategies for Global Change* 2010 (16): 119–142.

Chaturvedi, R.K., R. Gopalakrishnan, M. Jayaraman, G. Bala, N.V. Joshi, R. Sukumar, and N.H. Ravindranath. 2011. Impact of climate change on Indian forests: A dynamic vegetation modeling approach. *Mitigation and Adaptation Strategies for Global Change* 16 (2): 119–142.

Chazdon, R.L. 2008. Beyond deforestation: Restoring forests and ecosystem services on degraded lands. *Science* 320: 1458–1460.

Chhatre, A., and A. Agrawal. 2009. Trade-offs and synergies between carbon storage and livelihood benefits from forest commons. *Proceedings of the National Academy of Sciences* 106 (42): 17667–17670.

Cinner, J.M., B.P. Fuentes, and H. Randriamahazo. 2009c. Exploring social resilience in Madagascar's marine protected areas. Ecology and Society 14. http://www.ecologyandsociety.org/vol14/iss1/art41/.

Cutter, S.L., B. Boruff, and W.L. Shirley. 2001. Indicators of social vulnerability to hazards. Unpublished paper. Columbia, SC: University of South Carolina, Hazards Research Lab.

Cutter, S.L., B.J. Boruff, and W.L. Shirley. 2003. Social vulnerability to environmental hazards. *Social Science Quarterly* 84 (2): 243–261.

Cutter, S.L., J.T. Mitchell, and M.S. Scott. 2000. Revealing the Vulnerability of People and Places: A Case Study of Georgetown County, South Carolina. *Annals of the Association of American Geographers* 90 (4): 713–737.

Danda, A.A., A. Roy, and A. Choudhury. 2010. *Sundarbans: Future Imperfect Climate Adaptation Report*. World Wide Fund-India Report.

Danda, Anamitra Anurag, Gayathri Sriskanthan, Asish Ghosh, Jayanta Bandyopadhyay, and Sugata Hazra. 2011. *Indian Sundarbans delta: A vision*. New Delhi: World Wide Fund for Nature-India.

Deressa, T., R.M. Hassan, T. Alemu, M. Yesuf, and C. Ringler. 2008. *Analyzing the determinants of farmers' choice of adaptation methods and perceptions of climate change in the Nile Basin of Ethiopia*. International Food Policy Research Institute (IFPRI) Discussion Paper No. 00798. Washington DC: Environment and Production Technology Division, IFPRI.

Deressa, T.T., R.M. Hassan, and C. Ringler. 2009. *Assessing household vulnerability to climate change: The case of farmers in the Nile Basin of Ethiopia*. IFPRI Discussion Paper No. 00935, 18. Washington, DC: International Food Policy Research Institute.

Downing, T.E., and A. Patwardhan. 2004. *Vulnerability assessment for climate adaptation*. APF Technical Paper 3. New York City, NY: United Nations Development Programme. Final draft.

Du Toit, M.A., S. Prinsloo, and A. Marthinus. 2001. El Niño-southern oscillation effects on maize production in South Africa: A preliminary methodology study. In *Impacts of El Niño and Climate Variability on Agriculture*, eds. C. Rosenzweig, K.J. Boote, S. Hollinger, A. Iglesias, and J.G. Phillips, 77–86. ASA Special Publication 63. Madison, WI: American Society of Agronomy.

Easter, C. 1999. Small states development: A commonwealth vulnerability index. *The Round Table* 351 (1): 403–422.

Esty, D.C., M. Levy, T. Srebotnjak, and A. de Sherbinin. 2005. *2005 environmental sustainability index: Benchmarking national environmental stewardship*. IPH/Epi Reports No. 2004-009, D/2004/2505/17. New Haven, CT: Yale Center for Environmental Law & Policy Excess mortality in Belgium during the summer of 2003. Epidemiology section. Bruxelles, Institut Scientifique de Santé Publique.

FAO. 1991. Non wood forest products: The way ahead. Rome, Italy.

———. 2006. *Global forest resources assessment-progress towards sustainable forest management*. FAO Forestry Paper 147. Rome: Food and Agriculture Organisation of the United Nations.

FAO (Food and Agricultural Organization). 2010. *Managing forest for climate change*. I1960E/1/11.101960E.

FAO and SIDA. 1991. *Restoring the Balance. Women and Forest Resources*. Rome: Food and Agriculture Organization of the United Nations.

Fisher, D.R. 2001. Resource dependency and rural poverty: Rural areas in the United States and Japan. *Rural Sociology* 66: 181–202.

Fisher, M. 2004. Household welfare and forest dependence in Southern Malawi. *Environment and Development Economics* 9: 135–154.

Freudenberg, W.R., and S. Frickel. 1994. Digging deeper: Mining-dependent regions in historical perspective. *Rural Sociology* 59: 266–288.

Füssel, H. 2007a. Vulnerability: A generally applicable conceptual framework for climate change research. *Global Environmental Change* 17 (2): 155–167.

———. 2007b. Vulnerability: A generally applicable conceptual framework for climate change research. *Global Environmental Change* 17 (2): 155–167.

———. 2007c. Adaptation planning for climate change: Concepts, assessment approaches, and key lessons. *Sustainability Science* 2 (2): 265–275.

Füssel, H.-M., and R.J.T. Klein. 2006. Climate change vulnerability assessments: An evolution of conceptual thinking. *Climate Change* 75 (3): 301–329.

Gbetibouo, G.A. 2009. Understanding farmers' perceptions and adaptations to climate change and variability: The case of the Limpopo Basin, South Africa. Intl Food Policy Res Inst.

Ghosh, Aditya. (2012). Living with changing climate: Impact, vulnerability and adaptation challenges in Indian Sundarbans. Centre for Science and Environment. http://cseindia.org/userfiles/Living%20with%20changing%20climate%20report%20low%20res.pdf.

Godoy, R., N. Brokaw, and D. Wilkie. 1995. The effect of income on the extraction of non-timber tropical forest products: model, hypotheses and preliminary findings from the Sumu Indians of Nicaragua. *Human Ecology* 23: 29–52.

Grimes, A., B. Bennett, P. Jahnige, S. Loomis, M. Burnham, K. Onthank, W. Palacios, C. Cern, D. Neill, M. Balick, and R. Mendelsohn. 1994. The value of tropical forests: A study of non-timber forest products in the primary forest of the upper Napo province, Ecuador. *Ambio* 23: 405–410.

Hahn, M.B., A.M. Riederer, and S.O. Foster. 2009. The livelihood vulnerability index: A pragmatic approach to assessing risks from climate variability and change—A case study in Mozambique. *Global Environmental Change* 19 (1): 74–88. doi:10.1016/j.gloenvcha.2008.11.002.

Hair, J.F., W.C. Black, B.J. Babin, R.E. Anderson, and R.L. Tatham. 2006. *Multivariate Data Analysis*. 6th ed. Upper Saddle River, N.J.: Pearson Prentice Hall.

Hammill, P., R. Matthew, and E. Mc Carter. 2008. Microfinance and climate change adaptation. *IDS Bulletin* 39 (4): 113–122.

Hazra, S., Kaberi Samanta, Anirban Mukhopadhyay, and Anirban Akhand. 2010. *Temporal Change Detection (2001–2008) Study of Sundarban (Final Report)*. School of Oceanographic Studies Jadavpur University.

Heckman, J.J. 1976. The common structure of statistical models of truncation, sample selection and limited dependent variables and a simple estimator for such models. *Annals of Economic and Social Measurement* 5: 475–492.

Hedge, R., S. Suryaprakash, L. Achoth, and K.S. Bawa. 1996. Extraction of non-timber forest products in the forests of Biligiri Rangan Hills, India: Contribution to rural income. *Economic Botany* 50 (3): 243–251.

Hegde, P. 1998. Chipko and Appiko: How the people save the trees. Non-violence in Action Series; Quaker Peace and Service.

Heltberg, R., P.B. Siegel, and S.L. Jorgensen. 2009. Addressing human vulnerability to climate change: Toward a 'no-regrets' approach. *Global Environmental Change* 19 (1): 89–99.

Hobley, M. 1996. *Participatory Forestry: The Process of Change in India and Nepal*. London: Overseas Development Institute.

Howden, S.M., J. Soussana, F.N. Tubiello, N. Chhetri, M. Dunlop, and H. Meinke. 2007. Adapting Agriculture to Climate Change. *Proceedings of the National Academy of Sciences* 104: 19691–19696.

Humphrey, C.R. 1994. Introduction: Natural resource-dependent communities and persistent rural poverty in the U.S.—Part II. *Society and Natural Resources* 7: 201–203.

Huq, S., H. Reid, M. Konate, A. Rahman, Y. Sokona, and F. Crick. 2004. Mainstreaming adaptation to climate change in Least Developed Countries (LDCs). *Climate Policy* 4: 25–43.

IFAD. 2002. *Assessment of Rural Poverty: Asia and the Pacific*. Rome: International Fund for Agricultural Development.

INCCA. 2010. Climate Change and India: A 4X4 Assessment, Government of India.

Ionescu, C., R.J.T. Klein, J. Hinkel, K.S.K. Kumar, and R. Klein. 2009. Towards a formal framework of vulnerability to climate change. *Environmental Modeling and Assessment* 14 (1): 1–16.

IPCC. 2001a. *Climate Change Impacts (2001), Adaptation and Vulnerability*. Cambridge: Cambridge University Press.

———. 2001b. Climate change 2001: The scientific basis. In *Contribution of Working Group 1 to the Third Assessment Report of the Intergovernmental Panel on Climate Change*, ed. J.T. Houghton, Y. Ding, D.J. Griggs, M. Noguer, P.J. van der Linden, X. Dai, K. Maskell, and C.A. Johnson. Cambridge, UK: Cambridge University press.

———. 2001c. Climate change 2001: Impacts, adaptation and vulnerability. In *Contribution of Working Group 2 to the Third Assessment Report of the Intergovernmental Panel on Climate Change*, ed. J.J. McCarthy, O.F. Canziani, N.A. Leary, D.J. Dokken, and K.S. White. Cambridge, UK: Cambridge University Press.

———. 2007a. *Climate Change 2007—Impacts, Adaptation and Vulnerability—Contributions of Working Group II to the Fourth Assessment Report of the International Panel on Climate Change*. Cambridge University Press: Cambridge.

———. 2007b. Summary for policy makers. In *Climate Change 2007: The Physical Science Basis. Contribution of Working Group 1 to the Fourth assessment Report of the Intergovernmental Panel for Climate Change*, ed. S. Solomon, D. Qin, M. Manning, Z. Chen, M. Marquis, K.B. Averyt, M. Tignor, and H.L. Miller, 18. Cambridge: Cambridge University Press.

———. 2007c. *Climate Change 2007: Synthesis Report. Contribution of Working Group I, II and III to the Forth Assessment Report of the Intergovernmental Panel on Climate Change*, eds. Core writing team, R.K. Pachauri, and A. Reisinger, 104. Geneva, Swizerland: IPCC.

Jodha, N.S. 1986. Common property resources and rural poor in dry regions of India. *Economic and Political Weekly* 21(26).

———. 1992. Common property resources and rural poor in dry regions of India. *Economic and Political Weekly* 21: 1169–1181.

Johnson, T.G., and J.I. Stallman. 1994. Human capital investment in resource-dominated economies. *Society and Natural Resources* 7: 221–223.

Joshi, S. 2003. Super market, secretive. Exploitative, is the market in the minor forest produce unmanageable? *Down to Earth* 28: 27–34.

Kaiser, H.M., S.J. Riha, D.S. Wilks, D.G. Rossiter, and R.K. Sampath. 1993. A farm-level analysis of economic and agronomic impacts of gradual warming. *American Journal of Agricultural Economics* 75: 387–398.

Kaly, U., L. Briguglio, H. McLeod, S. Schmall, C. Pratt, and R. Pal. 1999. *Environmental vulnerability index (EVI) to summarise national environmental vulnerability profiles*. SOPAC Tech. Rep. 275. Suva, Fiji: South Pacific Applied Geoscience Commission.

Kaly, U., and C. Pratt. 2000. *Environmental vulnerability index: Development and provisional indices and profiles for Fiji, Samoa, Tuvalu and Vanuatu. Phase II. Report for NZODA*. SOPAC Technical Report 306.

Kavi Kumar, K.S., and J. Parikh. 2001. Socio-economic impacts of climate change on Indian agriculture. *International Review for Environmental Strategies* 2 (2): 277–293.

Kelly, P.M., and W.N. Adger. 2000. Theory and practice in assessing vulnerability to climate change and facilitating adaptation. *Climatic Change* 47: 325–352.

Kumar Kavi, K.S. 2002. Vulnerability and adaptation of agriculture and coastal resources.

Lamb, D., P.D. Erskine, and J.A. Parrotta. 2005a. Restoration of degraded tropical forest landscapes. *Science* 310: 1628–1632.

———. 2005b. Restoration of degraded tropical forest landscapes. *Science* 310: 1628–1632.

Lynam, T., R. Cunliffe, and I. Mapaure. 2004. Assessing the importance of woodland landscape locations for both local communities and conservation in Gorongosa and Muanza Districts, Sofala Province, Mozambique. *Ecology and Society* 9(4), art. 1.

Maddison, D. 2006. *The perception of and adaptation to climate change in Africa*. CEEPA Discussion Paper No. 10. Pretoria, South Africa: University of Pretoria, Centre for Environmental Economics and Policy in Africa.

Mahapatra, B., K. Sahad, and N.C. Dutta. 1993. *Destruction of Shellfish and Finfish Seed Resources of the Sundarbans, West Bengal and Suggestions for Their Conservation*. Ramkrishna Ashram Krishi Vigyan Kendra, Nimpith Ashram. South 24-Parganas, West Bengal, India.

Malhotra, K.C., Dutta, M., Vasulu, T.S., Yadav, G., and Adhikari, M. 1991. Role of NTFP in village economy: A household survey in Jamboni Range, Midnapore District, West Bengal, India, Institute of Bio-social Research and Development (IBARD), Calcutta, India.

Mamo, G., E. Sjaastad, and P. Vedeld. 2007. Economic dependence on forest resources: A case from Dendi District, Ethiopia. *Forest Policy and Economics* 9 (8): 916–927.

Marshall, N.A., and P.A. Marshall. 2007. Conceptualising and Operationalising Social Resilience within Commercial Fisheries in Northern Australia. Ecology and Society 12. http://www.ecol-ogyandsociety.org/vol12/iss11/art11/.

Marshall, N.A. 2008a. *A Conceptual and Operational Understanding of Social Resilience. Insights for Optimising Social and Environmental Outcomes in the Management of Queensland's Commercial Fishing Industry*. VDM Verlag: Saarbrücken Germany.

McCarthy, J.J., et al. 2001. *Climate Change 2001: Impacts, Adaptation, and Vulnerability, Contribution of Working Group II to the Third Assessment Report of the IPCC*. Cambridge: Cambridge University Press.

McClanahan, T., and Cinner J. 2009. Coastal communities responses to disturbance. In *Adapting to a Changing Environment: Global Warming Consequences and Proposed Actions for the Western Indian Ocean*. Oxford University Press.

Mendelsohn, R., W. Nordhaus, and D. Shaw. 1994. Measuring the impact of global warming on agriculture. *American Economic Review* 84: 753–771.

Mertz, O., K. Halsnæs, J.E. Olesen, and K. Rasmussen. 2009a. Adaptation to climate change in developing countries. *Environmental Management* 43: 743–752.

Milbrath, L.W. 1995. Psychological, cultural, and informational barriers to sustainability. *Journal of Social Issues* 51 (4): 101.

Milne, G. 2006. *Unlocking opportunities of forest dependent people in India*. Report No. 34481- in Volume 1 2006, World Bank.

Mishra, T.K., D. Mandal, and S.K. Maity. 2006. Evaluation of regeneration of *Shorea robusta* forests under Joint Forest Management in West Bengal. *International Journal of Environment and Sustainable Development* 5 (1): 12–22.

Mitchell, C.P., S.E. Corbirdge, S.L. Jewit, A.K. Mahapatra, and S. Kumar. 2003. *Non Timber Forest Products: Availability, Production, Consumption, Management and Marketing in Eastern India*. London: University of Aberdeen.

MoEF. 2006. *Report of the National Forest Commission*, 421. New Delhi: Ministry of Environment and Forests, Government of India.

———. 2009. *Climate Change Negotiations: India's Submissions to the United Nations Framework Convention on Climate Change*, 58. Delhi: Ministry of Environment and Forests, Government of India.

MoEF (Ministry of Environment and Forests). 2012. *India's Second National Communications to the United Nations Framework Convention on Climate Change*. New Delhi: Ministry of Environment and Forests.

Mukherjee S. 2001. Low quality migration in India: The phenomena of distressed migration and acute urban decay. *Paper presented at the 24th Iussp Conference*, Salvador, Brazil.

Nanda, Kumar P., and P.C. Sutar. 2001. Management of forest fire through local communities: A study in the Bolangir, Deogarh and Sundergarh Districts of Orissa, India. www.fao.org/DOCREP/006/AD348E/ad348e0g.htm.

Narayan, D., R. Chambers, M.K. Shah, and P. Pettesch. 2000. *The Voices of the Poor: Crying Out for Change*. Washington: World Bank.

NATCOM. 2004. India's Initial National Communication to the United Nations Framework convention on climate change, 268. Ministry of Environment and Forests. *Natural Resources* 8, 111–131.

Nelson, R., S.M. Howden, and Smith M. Stafford. 2008. Using adaptive governance to rethink the way science supports Australian drought policy. *Environmental Science & Policy* 11: 588–601.

Nhemachena, C., and R. Hassan. 2007. *Micro-level analysis of farmers' adaptation to climate change in Southern Africa*. IFPRI Discussion Paper No. 00714. Washington, DC: International Food Policy Research Institute.

Nord, M. 1994. Natural resources and persistent poverty: In search of the nexus. *Society and Natural Resources* 7: 205–220.

O'Brien, K., R. Leichenko, U. Kelkar, H. Venema, G. Aandahl, H. Tompkins, A. Javed, S. Bhadwal, S. Barg, L. Nygaard, and J. West. 2004. Mapping vulnerability to multiple stressors: Climate change and globalization in India. *Global Environmental Change* 14: 303–313.

Olsen, J.E., P.K. Bocher, and Y. Jensen. 2000. Comparison of scales of climate and soil data for aggregating simulated yields in winter wheat in Denmark. *Agriculture, Ecosystems and Environment* 82 (3): 213–228.

Overdevest, C., and G.P. Green. 1995. Forest dependence and community well-being: A segmented market approach. *Society and Natural Resources* 8: 111–131.

Palit, S. 1995. Role of NTFP in joint forest management. In *Proceedings of Seminar on Joint Forest Management (JFM)*. March 8–9, Calcutta. ed. M.R. Girish, 1998, PhD thesis.

Pandey, D.N. 2002. Sustainability science for tropical forests. *Conservation Ecology* 6(1). http://www.consecol.org/vol6/iss1/resp13.

Pantoja, E. 2002. Microfinance and Disaster Risk Management: Experiences and Lessons Learned' Prevention Consortium Draft Final Report (July 2002).

Parry, M.L., O.F. Canziani, J.P. Palutikof, P.J. van der Linden, and C.E. Hanson, eds. 2007. *Contribution Of Working Group II to the Fourth Assessment Report of the Intergovernmental Panel on Climate Change*. Cambridge: Cambridge University Press.

Pattanayak, S., V.P. Sharma, N.L. Kalra, V.S. Orlov, and R.S. Sharma. 1994. Malaria paradigms in India and control strategies. *Indian Journal of Malariology* 31 (4): 141–199.

Peluso, N.L., C.R. Humphrey, and L.P. Fortmann. 1994. The rock, the beach, and the tidal pool: People and poverty in natural resource-dependent areas. *Society and Natural Resources* 7: 23–38.

Pervez, M.S. 2002. Role of non-timber forest products in the economy of dwelling households of Dhading district, Nepal: An Economic Analysis. M.Sc. thesis, University of Agricultural sciences, Bangalore.

Piya, Luni, Keshav Lall Maharjan, and Niraj Prakash Joshi. 2012.Vulnerability of rural households to climate change and extremes: Analysis of Chepang households in the Mid-Hills of Nepal. *Selected Paper Prepared for Presentation at the International Association of Agricultural Economists (IAAE) Triennial Conference*, Foz do Iguaçu, Brazil, August 18–24.

Poggie, J.J., and C. Gersuny. 1974. Fishermen of galilee. The Human Ecology of a New England Coastal Community. National Oceanic and Atmospheric Administration, Rhode Island University Marine Bulletin Series Number 17 No., Kingston.

Polsky, C., and W.E. Esterling. 2001. Adaptation to climate variability and change in the US Great Plains: A multi-scale analysis of Ricardian climate sensitivities. *Agriculture, Ecosystems and Environment* 85 (3): 133–144.

Prasad, R., and P. Bhatanagar. 1993. Non-wood forest products and the indigenous fringe dwellers in Madhya Pradesh. *Journal of Tropical Forestry* 9: 188–195.

Prescott-Allen, Robert. 2001. *The Wellbeing of Nations: A Country-by-Country Index of Quality of Life and the Environment*. Washington, DC: Island Press.

Rao, A. Ratna, and B.P. Singh. 1996. Non-wood Forest products contribution in tribal economy. *Indian Forester* 122 (4): 337–342.

Ravindranath, N.H., et al. 2011. Climate change vulnerability profiles for North East India. *Current Science* 101(3).

Ravindranath, N.H., N.V. Joshi, R. Sukumar, and A. Saxena. 2006. Impact of climate change on forest in India. *Current Science* 90 (3): 354–361.

Rickson, R.E., J.S. Western, and R.J. Burdge. 1990. Social impact assessment: Knowledge and development. *Environmental Impact Assessment Review* 10: 1–10.

Roy, J., A. Ghosh, A. Majumdar, P. Roy, A.P. Mitra, and C. Sharma. 2005. Socio-economic and physical perspectives of water related vulnerability to climate change: Results of field study in India. *Science and Culture* 71 (8): 239–259.

Rygel, L., D. O'Sullivan, and B. Yarnal. 2006. A method for constructing a social vulnerability index: An application to hurricane storm surges in a developed country. *Mitigation and Adaptation Strategies for Global Change* 11 (3): 741–764.

Saha, A., and B. Guru. 2003. *Poverty in remote rural areas in India: A review of evidence and issues*. GIDR WORKING, 139:69. Ahmedabad: Gujarat Institute of Development Research.

Sarker, D.N., and N. Das. 2008. A study of economic outcome of joint forest management programme in West Bengal: The strategic decisions between government and forest fringe community. *Indian Economic Review* 43 (1): 17–45. Delhi.

Sarkar, S., and R.N. Padaria. 2010. Farmers' awareness and risk perception about climate change in coastal ecosystem of West Bengal. *Indian Research Journal of Extension Education* 10 (2): 32–38.

Sen, A. 1981. *Poverty and Famines*. Oxford: Oxford University Press.

Sharma, R.C. 2003. Forest for poverty alleviation: Chhattisgarh experience. In *Forests for Poverty Reduction: Changing Role for Research, Development and Training Institutions*, ed. H.C. Sim, S. Appanah, and N. Hooda. Bangkok: FAO.

Sharma, V.P., A. Srivastava, and B.N. Nepal. 1994. A study of the relationship of rice cultivation and annual parasite incidence of malaria in India. *Social Science and Medicine* 38: 165–178.

Shougong, Z., L. Weichang, L. Wenming, and Huafeng. 2003. Community forestry in mountain development: A case study in Guizhou Province, China. In *Forests for Poverty Reduction: Changing Role for Research, Development and Training Institutions*, ed. H.C. Sim, S. Appanah, and N. Hooda. Bangkok: FAO.

Shrestha, K.L., M.L. Shrestha, N.M. Shakya, M.L. Ghimira, and B.K. Sapkota. 2003. Climate change and water resources in Nepal. In *Climate Change and Water Resources in South Asia, Proceedings of Year-End Workshop*, Katmandu, 7–9, January, 2003, 135.

Singh, S.P., M. Manish, B.R. Phukan, and A. Mishra. 2007. Economics of forest fire management. In *Disaster Management in India: Perspectives, Issues and Strategies*, ed. N. Rai and A.K. Singh, 155–166. Lucknow: New York Book Company.

Smit, B., I. Burton, R. Klein, and J. Wandel. 2000. An anatomy of adaptation to climate change and variability. *Climatic Change* 45: 223–251.

Smit, B., and Johanna Wandel. 2005. Adaptation, adaptive capacity and vulnerability. *Global Environmental Change* 16: 282–292.

Smith, R.C., et al. 2003. Palmer long-term ecological research on the Antarctic marine ecosystem. In *Antarctic Peninsula Climate Variability: Historical and Paleoenvironmental Perspectives. Antarctic Research Series, 79*, ed. E. Domack et al., 131–144. Washington, DC: AGU.

SOPAC. 2005. *Building Resilience in SIDS: The Environmental Vulnerability Index*. Suva, Fiji: South Pacific Applied Geoscience Commission.

Sorenson, R.L., and K. Kaye. 1999. Conflict management strategies used in successful family businesses. *Family Business Review* 12: 133.

South Pacific Applied Geoscience Commission and United Nations Environment Programme SOPAC. 2005. *Building Resilience in SIDS: The Environmental Vulnerability Index*. Suva, Fiji: South Pacific Applied Geoscience Commission.

Springate-Baginski, O., and E. Wollenberg, eds. 2010. *REDD. Forest Governance and Rural Livelihoods: The Emerging Agenda*, 279. Bogor: CIFOR.

Srinivasan, J. 2012. Impacts of climate change on India. In *Handbook of Climate Change and India: Development, Politics and Governance*, ed. Navroz K. Dubash, 29–40. Abingdon: Earthscan.

Steinfeld, J.I. 2001. Climate change and energy options: Decision making in the midst of uncertainty. *Fuel Processing Technology* 71: 121–129.

Sunderland, T.C.H., L.E. Clark, and P. Vantomme. 1999. Non-Wood Forest Products of Central Africa: Current research issues and prospects for conservation and development. In: *Central African Regional Program for the Environment (CARPE)*, 2001. Rome: Food and Agriculture Organization.

Taknet, D.K. 2002. *The Heritage of Indian Tea: The Past, the Present and the Road Head*, 115–124. New Delhi, India: vedams e Books (P) Ltd.

Taylor, F.W., and N.T. Parratt. 1995. The potential of non-timber forest products of Botswana. *Paper Presented in the Sixth Conference on the Australasian Council on Tree and Nut Crops Inc.* (ACOTANC-1995), Lismore, NSW, Australia, 11–15 September 1995.

Trieu, V.H. 2003. Forestry for poverty reduction in Viet Nam. In *Forests for Poverty Reduction: Changing Role for Research, Development and Training Institutions*, ed. H.C. Sim, S. Appanah, and N. Hooda. Bangkok: FAO.

Tshering, D. 2003. Forests for poverty alleviation: Case of Bhutan. In *Forests for Poverty*.

UNDP. 1990. *Human Development Report 1990*. New York: Oxford University Press.

———. 2005. *Human Development Report*. New York: United Nations Development Program (UNDP).

Van de Ven, W.P.M.M., and B.M.S. van Praag. 1981. Risk aversion and deductibles in private health insurance: An application of an adjusted tobit model to family health care expenditures. In *Health, Economics, and Health Economics*, ed. J. van der Gaag and M. Perlman. Amsterdam: North-Holland.

Wills, Russel M., and Richard G. Lipsey. 1999. An economic strategy to develop Non-Timber Forest Products and Services in British Columbia Forest Renewal BC Project No. PA97538-ORE. Final Report. In: *Integrating Non-Timber Forest Products into Forest Planning and Practices in British Columbia Forest practice Board, 2004*.

Wollenberg, E., and B. Belcher. 2001. In European Tropical Forest Research Network (ETFRN news:32).

World Bank. 2002. *World Development Report 2003*. Washington, DC: World Bank.

———. 2006. *India: unlocking opportunities for forest dependent people in India*. Report 85:34481. South Asia Region. Washington, DC: World Bank.

Wu, S., B. Yarnal, and A. Fisher. 2002. Vulnerability of coastal communities to sea-level rise: A case study of Cape May County, New Jersey, USA. *Climate Research* 22.

Xiao, X., et al. 2002. *Transient climate change and potential croplands of the world in the 21st century*. Massachusetts Institute of Technology, Joint Program on the Science and Policy of Global Change, Report No. 18. Cambridge: MIT.

Yamin, F., A. Rahman, and S. Huq. 2005. Vulnerability, adaptation, and climate disasters: A conceptual overview. *IDS Bulletin* 36 (4): 1–14.

Yirga, C.T. 2007. The dynamics of soil degradation and incentives for optimal management in Central Highlands of Ethiopia. Ph.D. thesis, Department of Agricultural Economics, Extension and Rural Development, University of Pretoria, South Africa.